Agriculture in an Interdependent World: U.S. and Canadian Perspectives

by T. K. Warley

D1344922

Canadian-American Committee

Sponsored by
- C.D. Howe Research Institute (Canada)
- National Planning Association (U.S.A.)

Library of Congress Catalog Card Number 77-73919
ISBN 0-89068-040-X

C.D. Howe Research Institute (Montreal, Quebec) and
National Planning Association (Washington, D.C.)

Printed and bound in the U.S.A.
May 1977, $4.00

 C 439

CONTENTS

AGRICULTURE IN AN INTERDEPENDENT WORLD: U.S. AND CANADIAN PERSPECTIVES
by T. K. Warley

A STATEMENT BY THE CANADIAN-AMERICAN COMMITTEE TO ACCOMPANY THE REPORT ON

Agriculture in an Interdependent World: U.S. and Canadian Perspectives

After an interval of seven years, the Canadian-American Committee returns to the subject of U.S. and Canadian agriculture in a global setting. This remains one of the most important direct links between the natural and human resources of our two countries and the changing world around us.

Our last such study of this situation, prepared in 1970 by J. Price Gittinger, examined the commercial market opportunities over the long term for North American agricultural exports, the contribution of food aid to the needs of the Third World and possible Canada-U.S. cooperation in world wheat production and marketing.* That report necessarily reflected the circumstances at the close of the 1960s, particularly the surplus situation then existing in grains and other foodstuffs.

Although many of the central issues of those years continue to be important—indeed, some are of even greater concern today—the whole context of the debate on national and international policies for food and agriculture has been altered by a series of developments in the world economic and food environment, including:

▶ fundamental changes in world monetary and trading relationships;

▶ a deterioration in the world food situation;

▶ the rapid advance in world energy prices;

▶ the demands of the less-developed countries for a "new international economic order";

▶ and, accordingly, the acceptance by the developed countries of a greater responsibility for assisting the economic and agricultural development of the less-developed world;

▶ the apparent emergence of the nonmarket economies (most notably the Soviet Union) as continuing purchasers of irregular quantities of grains and protein foodstuffs; and

▶ the new round of trade negotiations in the GATT that has brought about changes in the relations among developed countries.

Among the consequences of these developments that emerge for North American agriculture are the following:

▶ the enhanced economic and diplomatic power that a productive agriculture confers on Canada and the United States;

---

*J. Price Gittinger, *North American Agriculture in a New World* (Washington, D.C. and Montreal: Canadian-American Committee, March 1970).

▶ the importance for North American agriculture of such matters as energy pricing, exchange rate changes, reserves policy, and the terms of access to both supplies and markets;

▶ the central importance of agricultural matters in relations between North America and each of the other groupings in the world—Western Europe and Japan, the nonmarket economies and the developing countries;

▶ the growing dependence of these regions (other than Western Europe) on North America's exportable food supplies, particularly for grains and animal feeds, and the importance to them of North American agriculture's productive capacity, stockpiling policies and export practices.

In view of these new realities and their profound implications for North American agriculture, the Canadian-American Committee decided that a new study was now appropriate. The author chosen for this assignment was T.K. Warley, Professor of Agricultural Economics at the University of Guelph. The Special Subcommittee on Agriculture, headed by Arden Burbidge, asked Professor Warley to examine U.S. and Canadian agricultural policies for dealing with world needs against the background of a changed international environment featuring the politicization of the world food problem, proposals on international commodity policies in quest of a new international economic order, and the treatment of agricultural products in the GATT multilateral trade negotiations.

As usual, members of the Canadian-American Committee, while availing themselves of the opportunity to comment on earlier drafts, have allowed the author full freedom to express his views. Without necessarily endorsing all of Professor Warley's conclusions, we find his paper to represent a well-written and researched analysis of a set of very complex and far-reaching issues. As such, we expect it to make a timely and significant contribution to better understanding of a situation important to us all. Consequently, we recommend that it be published as a Committee-sponsored report signed by its author.

FOOTNOTES TO THE STATEMENT

W.A. Strauss: Professor Warley has written a thoughtful paper. It is a good analysis and exposition of the place of agriculture and food supply in a changing world political environment. It states fairly the questions we face now and for the next several years about world food supply as those questions relate to growing population and rising expectations. There are two assumptions in the report, however, that may detract somewhat from the study results.

First, the author states that U.S. production of wheat over the next decade could reach about 70-80 million tons annually, an increase over the 50-million ton average in 1973–75, and most of this would be available for export. That total represents a significant increase over record production and seems implausible in the absence of ideal conditions.

Second, the close relationship between energy and agricultural production has not been probed extensively enough. Food supply can be significantly affected not only by the economics of energy, but also by its availability.

Apart from the effect of these two exceptions, the report should stir the debate of important policy issues that need airing and resolution on a scale beyond just Canada and the United States.

(Reference—Last sentence, page 8)
Arden Burbidge: While I agree that different policy emphasis on agricultural trade could be a source of Canadian-American conflict, it could also be pointed out that there could also be a conflict even if each country had the same policies. In my view, there is much less conflict potential between the United States and Canada in a sellers' market for agricultural products than in a buyers' market in which the United States would be in intensive competition for export business.

(Reference—Last paragraph, page 56)
Arden Burbidge: I would strongly support Professor Warley's conclusion here that any arrangement that "impairs efficiency in global agriculture resource use and diminishes the potential for North American agriculture" would result in serious damage to both domestic and international aspects of Canadian and American life and well-being.

MEMBERS OF THE CANADIAN-AMERICAN COMMITTEE SIGNING THE STATEMENT

Co-Chairmen

ROBERT M. MacINTOSH
Executive Vice-President, The Bank of Nova Scotia

RICHARD J. SCHMEELK
Partner, Salomon Brothers

Members

JOHN N. ABELL
Vice President and Director, Wood Gundy Limited

R.L. ADAMS
Executive Vice President, Continental Oil Company

J.D. ALLAN
President, The Steel Company of Canada, Limited

J.A. ARMSTRONG
President and Chief Executive Officer, Imperial Oil Limited

IAN A. BARCLAY
Chairman, British Columbia Forest Products, Limited

MICHEL BELANGER
President and Chief Executive Officer, Provincial Bank of Canada

ROY F. BENNETT
President and Chief Executive Officer, Ford Motor Company of Canada, Limited

ROD J. BILODEAU
Chairman of the Board, Honeywell Limited

ROBERT BLAIR
President and Chief Executive Officer, Alberta Gas Trunk Line Company Limited

J.E. BRENT
Director, IBM Canada Ltd.

PHILIP BRIGGS
Senior Vice President, Metropolitan Life Insurance Company

*ARDEN BURBIDGE
Park River, North Dakota

NICHOLAS J. CAMPBELL, JR.
New York, New York

SHIRLEY CARR
Executive Vice-President, Canadian Labour Congress

W.R. CLERIHUE
Executive Vice-President, Staff and Administration, Celanese Corporation

HON. JOHN V. CLYNE
MacMillan Bloedel Limited

STANTON R. COOK
President, Tribune Company

THOMAS E. COVEL
Marion, Massachusetts

GEORGE B. CURRIE
Vancouver, British Columbia

JOHN H. DICKEY
President, Nova Scotia Pulp Limited

JOHN S. DICKEY
President Emeritus and Bicentennial Professor of Public Affairs, Dartmouth College

WILLIAM DODGE
Ottawa, Ontario

WILLIAM EBERLE
President and Chief Executive Officer, Motor Vehicle Manufacturers Association of the United States

A.J. FISHER
President, Fiberglas Canada Limited

CHARLES F. FOGARTY
Chairman and Chief Executive Officer, Texasgulf Inc.

ROBERT M. FOWLER
President, C.D. Howe Research Institute

JOHN F. GALLAGHER
Vice President, International Operations, Sears, Roebuck and Company

W.D.H. GARDINER
Deputy Chairman & Executive Vice President, The Royal Bank of Canada

CARL J. GILBERT
Dover, Massachusetts

PAT GREATHOUSE
Vice President, International Union, UAW

A.D. HAMILTON
President and Chief Executive Officer, Domtar Limited

JOHN A. HANNAH
Executive Director, World Food Council

ROBERT H. HANSEN
Senior Vice President-International, Avon Products, Inc.

G.L. HARROLD
President, Alberta Wheat Pool

---

*See footnotes to the Statement.

R.A. IRWIN
Chairman, Consolidated-Bathurst Limited

ROBERT H. JONES
President, The Investors Group

DONALD P. KELLY
President and Chief Operating Officer, Esmark, Inc.

DAVID KIRK
Executive Secretary, The Canadian Federation of Agriculture

J.L. KUHN
President and General Manager, 3M Canada Limited

HERBERT H. LANK
Director, Du Pont of Canada Limited

FRANKLIN A. LINDSAY
Chairman, Itek Corporation

JULIEN MAJOR
Executive Vice-President, Canadian Labour Congress

FRANCIS L. MASON
Senior Vice President, The Chase Manhattan Bank

WILLIAM J. McDONOUGH
Executive Vice President, The First National Bank of Chicago

WILLIAM C.Y. McGREGOR
International Vice President, Brotherhood of Railway, Airline and Steamship Clerks

H. WALLACE MERRYMAN
Chairman and Chief Executive Officer, Avco Financial Services, Inc.

JOHN MILLER
President, National Planning Association

COLMAN M. MOCKLER, JR.
Chairman and President, The Gillette Company

DONALD R. MONTGOMERY
Secretary-Treasurer, Canadian Labour Congress

JOSEPH MORRIS
President, Canadian Labour Congress

THOMAS S. NICHOLS, SR.
Director, Olin Corporation

JOSEPH E. NOLAN
Tacoma, Washington

HON. VICTOR deB. OLAND
Halifax, Nova Scotia

CHARLES PERRAULT
President, Perconsult Ltd.

RICHARD H. PETERSON
Chairman of the Board, Pacific Gas and Electric Company

BEN L. ROUSE
Vice President-Business Machines Group, Burroughs Corporation

THOMAS W. RUSSELL, JR.
New York, New York

A.E. SAFARIAN
Dean, School of Graduate Studies, University of Toronto

W.B. SAUNDERS
Group Vice President, Cargill, Incorporated

ARTHUR R. SEDER, JR.
Chairman and President, American Natural Resources Company

A.R. SLOAN
President and General Manager, Continental Can International

R.W. SPARKS
President and Chief Executive Officer, Texaco Canada Limited

EDSON W. SPENCER
President and Chief Executive Officer, Honeywell Inc.

*W.A. STRAUSS
Chairman and President, Northern Natural Gas Company

ROBERT D. STUART, JR.
Chairman, The Quaker Oats Company

A. McC. SUTHERLAND
Senior Vice President, INCO Limited

WILLIAM I.M. TURNER, JR.
President and Chief Executive Officer, Consolidated-Bathurst Limited

W.O. TWAITS
Toronto, Ontario

MELVIN J. WERNER
Vice President, Farmers Union Grain Terminal Association

HENRY S. WINGATE
Formerly Chairman and Chief Officer, INCO Limited

FRANCIS G. WINSPEAR
Edmonton, Alberta

D. MICHAEL WINTON
Chairman, Pas Lumber Company Limited

WILLIAM S. WOODSIDE
President, American Can Company

ADAM H. ZIMMERMAN
Executive Vice President, Noranda Mines Limited

---

*See footnotes to the Statement.

Acknowledgments

The preparation of this essay entailed drawing upon the work of a great many persons who have written upon various aspects of the subjects treated in this monograph. The literature from which I have drawn was too large to cite in the text, but I should like to acknowledge my indebtedness to numerous intellectual creditors.

I should like to thank also the individuals and the many officials of governments, intergovernmental organizations and farm organizations in Europe and North America for the generous counsel they provided on aspects of the enquiry.

For their perceptive comments and suggestions, I am indebted to the following members of the Canadian-American Committee: Arden Burbidge (Chairman of the Agriculture Subcommittee), Raymond Davis, G.L. Harrold, David Kirk, William Kuhfuss, and W.B. Saunders.

Particular thanks are due to John Volpe and Sperry Lea, staff economists with the National Planning Association, for their unflagging enthusiasm and encouragement over an extended period and for their extremely helpful editorial assistance.

T.K. Warley
February 1, 1977

1

Introduction and Summary

This study has been undertaken at a time when international economic relations are being transformed radically and when the world community is faced with a range of new and intractable problems of great complexity. For most of the postwar period, the world economy was characterized by steady growth, a remarkable deepening of international interdependence and a broadening of international cooperation. The core premises of the international economic order seemed durable, and there was a broad consensus on the direction of its evolution. In addition, there was confidence that international economic institutions could cope with occasional disturbances and adapt to changing needs.

Toward the end of the 1960s, the prevailing mood of optimism began to erode as the postwar economic order started to exhibit inadequacies and to become increasingly crisis-prone. The international monetary system was unable to distribute equitably the benefits and burdens of a fixed exchange rate system, to effect orderly changes in parities and to cope with massive capital movements. The concept of an open, nondiscriminatory trading system was threatened by a swelling wave of protectionism, by the spread of discriminatory regional groupings and by the inability of the General Agreement on Tariffs and Trade (GATT) code of fair commercial practice to exert effective control over the trade distortions that flowed from the economic policies of governments. In a more pluralistic world economy, the new economic superpowers of Japan and a uniting Europe seemed unwilling to accept a share of responsibility for nurturing and reshaping the global economic order commensurate with their strength or their stake. The developing countries began to insist that their accelerated development be the prime objective rather than a by-product of world trade and monetary arrangements. The functional utility of multilateral institutions designed to handle aspects of international economic relations as discrete issues diminished as problems of trade, monetary cooperation and development became progressively more intertwined.

In the early 1970s, an already faltering world economic order was dealt a series of shocks of unprecedented variety and magnitude. The international monetary arrangements developed at Bretton Woods have been replaced abruptly by a system of floating exchange rates. The world has experienced a severe economic recession and rampant inflation from which it has only recently begun to recover. "Stagflation" has fed protectionist tendencies in commercial relations. The developing countries have moved from being supplicants to demanders of a radical reordering of world economic and political relations. The rapid advance in energy prices has led to important redistributions of political influence and wealth, and it has contributed to lower rates of real economic growth, higher levels of inflation and large monetary disturbances. A sharp deterioration in the world food situation has caused renewed concern about global finite limits. And, protection of the environment, access to private and common property rights, and transnational production are new issues that suddenly have sprung to the fore and commanded international attention.

These changes in the world economic and political environment are having multiple and interrelated effects on the agricultural industries of the United States and Canada. Realignments of exchange rates have intensified foreign demands on North American agriculture's productive capacity. Lower levels of real economic growth have slowed the rate of growth of demand for livestock products and feed grains that were previously the most dynamic demand centers. Inflation has shifted agricultural supply functions upward, made farmers more vulnerable to product price variations and enhanced consumers' interests in agricultural and food trade policies. Although energy price increases have raised the supply price of farm products and strengthened the importance of agricultural exports in the external accounts of the United States and Canada, they have eroded the ability of food deficit countries to expand their reliance on food imports from areas with exportable surpluses. The proposals of the developing countries for a new international economic order to govern their relations with the advanced countries have a direct bearing on the problem of world hunger, on the ability of the less-developed countries (LDCs) to provide growing markets for North American agriculture, on the competitive relations in world markets between North American farmers and LDC producers of similar products, and on the fundamental character of the economic system under which primary products are produced, priced and exchanged. High and unstable food import bills have enhanced the attractiveness to Europe and Japan of autarkic food policies and managed world food markets, and they have heightened the long-standing conflict among the developed countries on appropriate trade policies for agricultural products. The emergence of the USSR as a regular purchaser of uncertain quantities of Western grains has presented a valuable additional market for North American agriculture, but it has also introduced further instabilities into the world food system and posed difficult issues concerning how this component of world food trade should be handled.

The foregoing serves to point up five important characteristics of the place of agriculture in contemporary international economic relations.

▶ First, in a world in which international economic issues increasingly are intertwined, problems pertaining to food and agriculture cannot be viewed as a discrete problem set, nor will they yield to national and multilateral policies with a narrow agricultural focus.

▶ Second, no sector better demonstrates the reality of the deepening interdependence of the world economy and the corollary that the world's agricultural and food problems (freedom from hunger, security of food supply, efficient use of the world's agricultural resources) are but part of a growing list of transnational problems requiring international cooperation for their solution.

▶ Third, agricultural matters are of central importance in the economic and political relations between North America and each of the other major politico-economic groupings of countries with which the United States and Canada must deal—the less-developed countries, the members of the Organization of Petroleum Exporting Countries (OPEC), the socialist countries, Japan, and other advanced Western societies.

▶ Fourth, although there unquestionably has been a decline in the hegemony in international affairs of the United States in recent years, the area of food and agriculture serves to re-emphasize the importance of its leadership, for there can be no solutions to the global issues of food supply, security and future trading relationships without its concurrence, and on most issues the rest of the world looks to the United States to take initiatives.

▶ Fifth, the strength of the North American economy, and thereby the ability of the United States and Canada to influence world economic and political events, is importantly dependent on the performance of their agricultural and food sectors in the international economy. Hence, the nature of the national and international arrangements made for agriculture in a rapidly changing world have a direct bearing on the economic well-being of the United States and Canada and on their position in world economic and political affairs.

These matters constitute the background for the present study. Not all of them can be explored directly or in depth in a study of modest length. The core of the study is in Chapters 3 to 6 that examine:

▶ the response of the United States and Canada and the contribution of their agricultural industries to the problem of world hunger and food insecurity;

▶ the effects on North American agriculture of the proposals made by the LDCs for a new strategy for dealing with problems of commodity trade;

▶ the future arrangements for agricultural products trade between the developed countries now being explored in the GATT multilateral trade negotiations; and

▶ the problems associated with the participation of the USSR in world food markets.

Chapter 2 provides a brief profile of North American agriculture, and Chapter 7 concludes by explaining how the changing external environment faced by North American agriculture affects the national agricultural trade, food and farm policies of the United States and Canada.

Throughout the study, an attempt is made to identify the implications for the United States and Canada of each of the topics listed above and their responses to the opportunities and problems presented. Particular attention is paid to the implications for the bilateral relationship associated with the different circumstances and policy priorities of each country.

SUMMARY

Chapter 2: A Profile of North American Agriculture

Changes in the structure of farming, the full integration of the agricultural and food system into the national economies of the United States and Canada and changes in farmers' prices and costs since 1972 have caused agricultural policy to be redirected in each country. In the United States, the role of government in farming generally has been reduced, but interventions to stabilize prices to consumers have become more common. In Canada, measures to protect farmers against fluctuations in their prices and costs have been extended. The differing policy responses to the problems of instability in the agricultural and food system in each country created recent bilateral trade conflicts, and the scope for these conflicts may well expand in the future.

Secular changes that have occurred in the trade of each country have increased the importance to their national economies of maintaining a high level of agricultural exports and thus of foreign agricultural trade policies. Structural changes in world grain trade have made other regions of the world more dependent on North America's grain exports, creating a possible additional source of diplomatic influence.

The aggressive use of "food power" is a concept lacking plausibility and appeal. However, agriculture makes an important contribution to the diplomatic influence of the United States and Canada by contributing to strength in their national economies.

Foreign countries need not be unduly concerned about North American agriculture's future production and export capacity. Although Canada's ability to expand food exports is limited by its resource endowments, the United States has a considerable capacity to expand food supplies. Furthermore, overseas buyers do not face an inexorable rise in the supply price of U.S. food exports.

Chapter 3: World Hunger and Food Insecurity

The world food system is functioning badly. World food supplies are inadequate, and malnutrition is widespread in many developing countries. The relief of hunger is primarily the responsibility of the governments of the LDCs. The physical capacity for expanding their indigenous food production is very large, but increasing the rate of growth of output requires that they reorder national priorities, increase investments in agriculture and rural infrastructure and provide their farmers with better economic incentives. In addition, it is imperative that an early start be made in reducing the rate of growth of their populations if per capita food supplies are to be increased.

The responsibility for helping the LDCs solve their food supply problems is shared by Western countries, OPEC and the centrally planned economies. The requirements are an expansion of agricultural development assistance, the provision of food aid, enhancement of world food security, and changes in trading arrangements in ways that would enable the LDCs to make wider use of world markets in their food supply strategies.

The United States and Canada have made an appropriate response in three of these areas. Their development assistance for agriculture has expanded and has been redirected toward the countries with the most severe food problems. Together they are providing 70 percent of total world food aid. The United States has exercised leadership in making a concrete proposal for the creation of a system of world food reserves to permit the maintenance of consumption in years of poor harvests. This proposal has met with a disappointing response from other countries, including Canada.

Little has been done in the area of trade policy changes that would enable the LDCs to increase their foreign exchange earnings and hence their ability to import more food. Additional measures are required that will stabilize world food supplies and prices and thereby reduce the risks the LDCs face in increasing their dependence on world markets for part of their growing food needs. The agricultural industries of Canada and the United States have a very large stake in measures that would encourage the LDCs to make more use of the trade option, since the developing countries represent a large and dynamic market for their grain exports.

Chapter 4: North American Agriculture and International Commodity Policy

Within the context of securing fundamental changes in all aspects of the economic relationships between rich and poor countries, the

LDCs have proposed a comprehensive strategy for dealing with the special problems they face in their commodity trade.

Detailed examination of the integrated program for commodities suggests that several of its elements are consistent with the trade principles and economic interests of the United States and Canada. These include liberalization of trade in products of interest to the LDCs and improved compensatory finance arrangements. However, other proposals are in conflict with the foreign economic policies of the United States. These include widespread resort to formal intergovernmental commodity agreements and their common funding, indexation of the prices of LDC commodity exports and the involvement of governments in supply and purchase commitments. Canada's position on these latter matters is not so clearly defined.

The grain sector of North American agriculture has much to gain from the introduction of measures that would accelerate LDC economic growth rates and stabilize and raise the developing countries' foreign exchange earnings. However, other sectors of North American agriculture that presently enjoy high rates of protection from LDC exports would be disadvantaged by trade liberalization for competitive products. Intensified competition would be experienced in both domestic and overseas markets. Furthermore, there are long-run dangers to the performance and the potential of North American agriculture, and to the world trading system, in the creation of an excessively regulated regime governing commodity production, pricing and trade.

Chapter 5: Agriculture in the GATT Negotiations

The current GATT negotiations offer an opportunity to secure both freer and fairer conditions for trade in temperate-zone products between the developed countries. In addition, many of the participating countries attach importance to the further objective of creating arrangements that would ensure greater stability in world markets.

Countries differ greatly in the weight they attach to each of the above objectives. The United States is concerned primarily with securing improved conditions of access for its exports of grains and oilseeds to the European and Japanese markets. Canada shares this objective but also seeks measures that would guard against collapse and disorder in the world wheat market. These might be provided by an international commodity agreement for grains such as that proposed by the European Community (EC). The United States has regarded enhanced market stability as a subordinate objective to improved access.

It must be assumed that other countries will seek improved access for their exports to the U.S. and Canadian markets. Dairy products and meats are the two major commodity groups in which each country has much to offer. A number of techniques are available for lowering the level of agricultural trade protection and providing expanded market opportunities for low-cost exporters of farm products.

Strengthening the GATT code of trade conduct as it applies to agriculture involves bringing export controls, safeguard procedures and national subsidy policies under more effective international surveillance and control. There are a variety of ways in which the GATT's rules on these matters might be tightened.

The prognosis for achieving significant liberalization of trade in agricultural products is not good. It would be a mistake for the United States and Canada to overreact if this should prove to be the case, for there are numerous other benefits to be secured from the negotiations as a whole.

Chapter 6: Trading with the Socialist Countries

The expansion of agricultural exports to the socialist countries is important to North America for both economic and political reasons. Although exchanges with China are not expected to grow, those with the USSR could expand significantly. However, the variability and unpredictability of the USSR's grain import requirements during this decade have become a destabilizing influence on world and national food markets.

Although the existence of the Canadian Wheat Board has provided some advantages for Canada in dealing with monopsonistic buying agencies, there is no reason to believe that a centralized system of grain marketing would be better suited to the circumstances of the United States than the free enterprise system that now operates. And, the problems that country faces in dealing with the USSR have been largely resolved by the bilateral intergovernmental grains agreement of 1975.

The agreement may have the effect of rendering Canada's grain trade with the USSR less stable. More importantly, there is now less reason for the USSR to participate in the multilateral arrangement on grain reserves that the United States has proposed, and this may prejudice the conclusion of such an arrangement. Hence, U.S.-USSR trade in grains may have been stabilized at the expense of greater supply and price instability for grains in the world as a whole.

Chapter 7: The Nexus of Farm, Food and Trade Policies

Until recently, public policies for food and agriculture in the United States and Canada were concerned primarily with raising and/or stabilizing commodity prices and farmers' incomes. However, changed circumstances have broadened the range of issues with which policy makers must deal. Changes in the world food situation and in economic relations as they pertain to agriculture between North America and the rest of the world have meant that foreign agricultural trade policies, and foreign economic policies generally, have been elevated in relative importance. Domestically, preoccupation with rapid inflation and deficits in national trade accounts has meant that more attention has been paid to the contribution of the agricultural and food system to national economic policy objectives.

The term "food policy" has gained currency. It signifies that policy for agriculture and food must now be multidimensional in its objectives and constituencies. More especially, it means that farm policy can no longer be considered in isolation but must be treated as a component of macroeconomic and foreign economic policy.

This has not required new programs and the introduction of new policy instruments. But, it has intensified concern with the balance between the interests of different groups in society (notably farmers and consumers), expanded the range of competitive objectives between which trade-offs must be made (e.g., export receipts and domestic price stability), and multiplied the number of government agencies that participate in the policy-making process and have stakes in the decisions that are made.

Though the process of formulating a coherent food policy is not completed in either Canada or the United States, the objectives are clear. One is to devise policies that will maximize the influence of each country in international affairs generally and, on matters pertaining to agriculture, ensure that the agricultural and food system contributes to national economic growth, internal stability and external payments balance. The other objective is to provide food prices for consumers that are consistent with a stabilized economy while also providing returns to farmers that will ensure a steady expansion of their output.

In the international sphere, the most important differences in trade policy between Canada and the United States have been the contrasting views on the role of an international grains arrangement with price-fixing provisions, and the willingness of the United States to take initiatives and the preference of Canada to await the outcome of debate between others on a range of consequential matters. Domestically, farm and food policies are taking quite different courses. U.S. authorities seem prepared to tolerate a good deal of instability in farm prices but stand ready to mitigate increases in consumer food prices. However, in Canada, consumers' interests do not appear to be given nearly as much weight as the interest of farmers in assurance against risks in unstable markets. It is this different policy emphasis in the two countries that creates the potential for expanding bilateral agriculture trade disputes between them.

2

A Profile of North American Agriculture

For much of the period since World War II, the prevailing view of farming in the United States and Canada was that it constituted a weak and declining sector of the North American economy and was beset by endemic economic problems that could be kept within socially tolerable bounds only by extensive public support programs.

The conventional analysis of the economics of farming led to several pessimistic conclusions. Three factors—the characteristics of the supply of and demand for farm products, the economic organization of the industry and the conditions in foreign markets—were seen as combining to produce continual disequilibrium in the industry's product and factor markets and unsatisfactory economic and social conditions for its participants. On the supply side, a continual flow of production-increasing and labor-displacing new technologies caused farm output to increase faster than effective domestic and foreign demand. The growth of domestic consumption was constrained by slow rates of population increase and by low and falling income and price elasticities of demand. Important livestock industries were thought to be internationally uncompetitive. The international comparative advantage of the crop sector of North American agriculture and the ability to develop foreign markets for output greater than domestic requirements were frustrated and limited by a lack of purchasing power in food deficit developing countries, by the autarkic food policies of the centrally planned economies, and by the protectionist agrarian policies of food-importing developed countries. World markets were oversupplied, unstable and unrewarding. The joint influence of these characteristics and circumstances created a chronic excess capacity in North American agriculture; a secular deterioration in its terms of trade; low and unstable returns to labor, capital and managerial resources engaged in farming; a continual contraction in the employment capacity of the industry; and difficult adjustment problems for rural communities.

It was recognized that there were some gains to be derived from this situation. The release of labor from farming was an important source of national economic growth. Domestic consumers spent a declining share of their rising incomes on food purchases, and food prices were remarkably stable. Foreign countries were able to buy the agricultural exports of North America at bargain prices, and accumulated grain surpluses were available to cushion fluctuations in world output and to provide food aid to needy people around the world. However, North America's farmers did not share proportionately in their rising productivity, and they and taxpayers had to carry a disproportionate share of the costs of adjusting to a global imbalance between agricultural output and effective demand.

Although invariably overstated, there was substantial validity in this perception of the economic characteristics of agriculture in the postwar years. Certainly the economic forces affecting North American agriculture were sufficiently strong and sustained to bring about a radical restructuring of the industry during the 1950s and 1960s, and the economic and social circumstances of farm people and rural communities were judged to be so unsatisfactory that successive U.S. and Canadian governments became continually and deeply involved in farm income support and stabilization programs.

STRUCTURAL ADAPTATION

The principal structural changes that have occurred in U.S. and Canadian farming over the postwar years are indicated by the data in Table 1 (see page 21). There has been a shrinkage in the number of farm businesses and a corresponding growth in the size of those remaining in terms of acreage, sales and assets per farm. A sharp contraction in the numbers of people engaged in primary agricultural production has been made possible by a massive growth in capital employed in the aggregate and per farm. With rising output, a declining labor force and more capital per person employed, productivity of the farm labor force in both countries has grown more rapidly than in their economies as a whole. The smaller number of farm operators has achieved rising incomes from farming operations.

However, the raw data do not tell the full story of the far-reaching changes that have occurred within agriculture and between farming and the rest of the economy. First, the decline in the number of census farms and their operators does not reveal the heterogeneity that has developed in farming over the last quarter of a century. Farm firms now comprise two distinct groups. At one extreme is a commercial sector consisting of a small number of relatively large farming businesses that yield their operators satisfactory levels of disposable incomes and rates of return to the resources employed at least as high as those earned by similar resources in other occupations. These are the 20 and 30 percent of all census farms that accounted for 77 and 70 percent of total farm production in the United States and Canada, respectively, in 1971. A proportion of the occupiers of the remaining census farms are engaged full-time in farming, but the majority of them have adjusted their labor inputs to the land and capital resources they command by becoming part-time farmers. Though income from farming in this group may be low, their total family income is generally satisfactory.

The second important change that has occurred is that farming has become functionally and economically integrated more fully into the rest of the economy. Marketing activities are now performed almost entirely by food processing and distribution industries. Expansion of output has been accomplished by using inputs purchased from the motor vehicle, energy, farm machinery, and agrochemical industries, with a corresponding decline in importance of the contribution of farmer-owned and farmer-provided inputs such as land and labor. Also, there has developed an integrated agricultural and food system consisting of input supply, farming and food processing and distribution industries.

Third, though declining in relative terms, the agricultural and food system continues to be a large and growing sector in the economies of both the United States and Canada. The food industry as a whole accounts for fully one-sixth of national product and one-fifth of employment in each country, and both output and employment are expanding. Food purchases still absorb close to one-sixth and one-fifth of total consumer expenditures in the United States and Canada, respectively, and food items have a weight of about 25 percent in each country's consumer price index. In addition, exports of agricultural origin comprise one-fifth and one-eighth of the total merchandise exports of the United States and Canada, respectively, and the agricultural and food system is one of the few sectors in which each country has a positive and growing surplus in its trade in goods with other countries.

Accordingly, the rates of growth of productivity, output, employment, and exports in the agricultural and food systems of the United States and Canada are of great importance to the overall performance of their economies in terms of the growth of national product and employment, the stability of prices, the distribution of income, and the achievement of external payments balance. By the same token, the well-being of all participants in the North American agricultural and food system is directly and increasingly dependent on the growth and stability of the national and international economies.

AGRICULTURAL POLICIES PRIOR TO 1972

The major thrusts of public policy for agriculture in both countries in the period from the end of World War II until the early 1970s were to redistribute income to farmers, to enhance market stability and to ameliorate the pace of farm and rural adjustment. Policies for food and agriculture were designed primarily to serve the needs of farmers. Consumers' interests were not given much attention, nor was this urgent because agricultural support and stabilization policies were never generous enough to halt the downward trend of real food prices. For most of the period, agricultural policies tended to be an agglomeration of commodity programs dealing with production levels, marketing arrangements and farm gate prices. Little attention was paid to the disparate circumstances of producers within an increasingly heterogeneous farming industry, to the emergence of an industrialized farming sector as a component of an integrated and expanding food system, or to the reciprocal relationships between the performance of the food system and the larger national economy. The growing importance of the North American agricultural and food system in an increasingly interdependent world economy and in international economic and political relations was but poorly perceived.

These characteristics became less pronounced as time passed, and there were important differences in the form, emphasis and evolutionary direction of the domestic farm programs and foreign agricultural trade policies of the United States and Canada in the period prior to 1972. The more far-reaching changes were in the United States. Beginning in the early 1960s, the development of U.S. farm policy had two dominant themes. The first was to expand the influence of market forces on resource use and incomes in agriculture. The second theme was to place more emphasis on expanding exports of farm products in order to release agriculture's full production capacity, to generate rising farm incomes, to expand foreign exchange earnings, and to reduce the level of public expenditure needed to support the industry. Over the years, these two themes merged, and domestic farm programs became progressively more consistent with an expansionist foreign trade policy.

In the previous decade, the support of product prices at too high levels for too long had led to the accumulation of huge government-owned stocks and necessitated the use of export subsidies, import restrictions, foreign surplus disposal operations, and rigid production controls. In the 1960s, support prices for the major export crops were lowered toward world market levels, and high loan rates were replaced by direct payments to farmers to maintain income and to induce them to participate in supply management programs. Producers who participated in the program to idle land were given progressively more latitude in the mix of crops they might produce on the acreage they planted. This redirection of U.S. domestic farm programs led eventually to the Agricultural and Consumer Protection Act of 1973, under which loan rates for grains and upland cotton were set at very low levels while deficiency payments were introduced as the technique for implementing minimum product price guarantees.

However, it may be noted that the concept of a "market-oriented agriculture" was selectively applied only to the major export crops and only with limited success. As late as 1972, 60 million acres of cropland were being held out of production, government-owned stocks of grains totaled 40 million metric tons (mmt.), and price support operations for grains and cotton were costing approximately $3 billion per year. High price supports, production and import controls and export assistance survived beyond 1972 for rice, peanuts and tobacco. Support for the milk industry remained extensive, and the beef and sheep industries were sheltered from international competition by a restrictive import policy for red meats.

Though the problems of excess capacity and depressed and unstable resource returns also plagued Canadian agriculture in the quarter century following World War II, the domestic farm programs of Canada differed from those of the United States in four main ways.

First, there was no parallel in Canada to the Pauline conversion of the United States to the importance of developing export markets for food and feed grains. The small absorptive capacity of the domestic market for grains relative to national production, the historic dependence of the economies of the Prairie provinces on grain exports, and the prominence of wheat exports in Canada's external accounts left no alternative to the vigorous overseas marketing of competitively priced grains as the cornerstone of national grains policy.

Second, the Canadian commitment to support farm product prices was more muted. Price and income objectives for agriculture were generally less ambitious and public expenditures on them more modest. Market forces were permitted considerable influence in bringing about adjustments within the industry. Indeed, the objective of agricultural price policies generally was limited to that of price stabilization. This was effected by a variety of measures, including a system of stop-loss commodity price guarantees set well below market-determined levels, participation in successive international wheat agreements and the operations of numerous provincial commodity marketing boards.

Third, Canada did not employ a comprehensive program of agricultural supply management. Though both federal and provincial policy arrangements for some commodities included supply management features, there was no general policy of production and marketing control through land withdrawal and publicly financed stock holding comparable to that which existed in the United States in the period prior to 1972.

Fourth, farm policy in Canada dealt somewhat more evenhandedly with both the crop and livestock sectors of agriculture. Most livestock products were included under federal price stabilization programs and the production, marketing and pricing of hogs, broilers, turkeys, and eggs all were influenced by numerous provincial marketing boards. Here again, Canadian farm policy differed from that of the United States where these commodities generally received only such support as was provided by the availability of ample supplies of feed grains at stable prices.

The foregoing is not meant to imply that Canadian agriculture was entirely unsupported in the 1950s and 1960s, for this was not the case.

Canadian milk producers, like their U.S. counterparts, were subsidized heavily by virtually every protective instrument known to policy makers. Beef producers received some protection from competition with lower-cost foreign suppliers, and Prairie grain growers were the recipients of significant benefits from public expenditures on grain storage, transport subsidies, foreign food aid, and, in 1970, by an emergency program to idle wheat land. Nevertheless, the general configuration of Canadian domestic agricultural policy throughout the postwar years was notably less interventionist than that of the United States, and, although it was more fragmented, agricultural policy was subject to less extensive changes in philosophy and program detail over the years.

In fact, it might be claimed that in shifting the objective of agricultural programs from income support to price stabilization, in embracing export market development as the key to agricultural development, in seeking to minimize public expenditures on agriculture, and in giving more rein to market forces to shape resource use and income distribution within agriculture and between agriculture and the rest of the economy, U.S. agricultural policy in the period 1961–72 gradually assumed more of the features that characterized farm policy in Canada throughout the postwar years.

For the most part, relations on agricultural trade matters between the United States and Canada were remarkably harmonious in the period up to 1972, despite differences in their respective farm programs. In bilateral exchanges of agricultural products, formal trade barriers between the two countries and the nontariff trade distortions induced by national farm programs were either modest (e.g., feed grains and oilseeds; live animals and red meats; poultry meats and eggs; fruits and vegetables) or where they were high (e.g., manufactured dairy products and wheat), they were consistent with the separate but similar policy objectives of each country. In extracontinental trade, Canada often had occasion to be concerned with the surplus disposal and export subsidy programs of the United States for wheat, but the worst excesses were avoided in competitive rivalry for third-country markets. Moreover, the dominant relationship in external agricultural relations was one of close cooperation in stabilizing world wheat markets and a coordinated approach to persuading Europe and Japan to grant improved access to their markets and to accept a larger measure of responsibility for production restraint, stock holding and the provision of food aid to developing countries.

The convergent tendencies in domestic agricultural policies and the absence of conflict in bilateral and external trade relationships in the period to 1972 are worth noting, for, as will emerge below, the divergent policy responses in the United States and Canada to the new environment for agriculture that has existed since that year have given rise to specific bilateral trade difficulties and to different approaches to addressing the agricultural and food problems of the future at both domestic and international levels.

POLICY DEVELOPMENTS
SINCE 1972

The exceptional level of world export demand for grains experienced since 1972 has brought to an end (temporarily at least) the long period of excess capacity in the crop sector of North American agriculture. Stocks of grain accumulated in earlier years have been moved into consumption, and the 60 million acres of cropland in the United States and the 10 million acres of wheat land in Canada that were being idled in 1971 have been returned to production. The surge in aggregate gross and net farm incomes experienced in both countries since 1972 has placed average returns to resources in agriculture fully on a par with those in other sectors of the economy. In addition, farmers have secured large capital gains from appreciation in land values, and the net exodus of people from farming has slowed to a trickle.

However, this general picture masks disparate tendencies within agriculture and new problems faced by the farming community as a whole. There has been a massive redistribution of income within farming. Producers of grains and oilseeds and other crops that compete for land have experienced huge increases in incomes and net worths. In contrast, the profitability of producing milk, meats and eggs has been eroded by a sharp increase in feed costs and by the weakening of demand for livestock products in the recession that began in 1973. Producers of most crops have been able to absorb the price increases of all farm inputs arising from general inflation and discontinuities in the market prices of energy, fertilizers and agrochemicals. Such cost increases have added further to the economic distress of producers of livestock products.

The effects on the nonfarm sectors of the U.S. and Canadian economies also have been uneven. Within the food system, input supply and marketing industries related to the crops sector have faired well, but those oriented toward animal agriculture have suffered economic dislocations. Sharp increases in retail food prices have exaggerated already established inflationary tendencies in each country and make the restoration of price stability more difficult. The historical tendency of expenditures on food to fall as a proportion of consumers' total outlays has been reversed, with adverse effects on the level and distribution of real incomes among different socioeconomic groups within society. On the positive side, a dramatic increase in the value of earnings from exports of agricultural products has been one of the strongest elements in each country's external accounts. The large surplus on agricultural trade limited the size of current account deficits, and thereby the depth of the recession, and permitted expansionary economic policies to be adopted at an earlier date.

There were few precedents to guide policy makers in both countries in their responses to the new issues raised by the abrupt changes in the circumstances of farmers and the new situations in national and international food markets that arose after 1972. Therefore, it is not surprising that some measures adopted in the short term have been in conflict with each other and with longer-term objectives. In addition, although there has been a discernible pattern in the direction of farm policy adaptation in each country, coherent farm and food policies for the future are far from being fully defined.

Four considerations have been especially influential in molding farm and food policy in the United States and Canada during the past several years. First, price and income instability has tended to replace chronic income inadequacy as the primary justification for public policies for agriculture. Second, there has been a movement toward viewing public policies for agriculture in the wider context of macroeconomic policy objectives. This has entailed elevating consumer interest in farm and food policies and increasing concern with the influence of food prices on inflationary tendencies and on the distribution of national income. Third, maintaining a high level of earnings from farm product exports has come to be perceived as essential to financing the rising import costs of energy and raw materials, to achieving balance in external payments and to sustaining the value of the dollar. Finally, the growing dependence of the rest of the world on North America's exportable supplies of basic foodstuffs is now seen as an important new feature of international economic relations.

In the United States, the long-held desire to reduce both the degree of government involvement in agriculture and public expenditures on farm programs has been made possible by circumstances. Loan rates and target prices for the major export crops have been set at levels well below market prices. Production controls for grains have been suspended, and the responsibility for holding stocks of grains and other commodities has been shifted to the private sector. Expenditures on agricultural price support programs have been greatly reduced. Attempts by Congress to raise crop price support levels closer to sharply higher production costs and to get the government back into stock holding have been opposed by the executive branch. Little relief has been provided for animal agriculture over a four-year period of severe economic stress. The support price for milk has been raised, but not by enough to prevent a deterioration of the ratio between milk prices and feed and other costs. Import regulations on red meats were tightened, but reduced profitability in the beef industry has forced a large contraction of the national beef herd. Producers of hogs, poultry and eggs and most fruits and vegetables have been provided with no protection against market instabilities and increased production costs. In general, U.S. authorities have shown a willingness to tolerate a high degree of instability in the farming component of the food system and especially in the livestock sector. However, concern with the economic and social consequences of general and food price inflation has precluded adherence to a completely *laissez-faire, laissez-passer* regime for food and agriculture. Interventions have been necessary, and these generally have favored domestic consumers. A short-lived attempt at direct price controls, a dramatic rise in consumer food subsidies, reductions in food aid shipments, sporadic export controls, and formal and informal bilateral export volume agreements with foreign buyers of U.S. grains and oilseeds have all been employed in the period 1972–76 in an attempt to hold down retail food prices.

Some developments in Canadian farm and food policies since 1972 parallel those in the United States. For instance, maximum grain production and exports have been encouraged, grain stocks have been moved into consumption, food aid shipments have been curtailed, and consumer prices of bread and some milk products have been subsidized. However, in one crucial respect the development of agricultural policy in Canada since 1972 has been diametrically opposite to that in the United States—the involvement of the federal and provincial governments in the affairs of the industry has expanded significantly. Several factors have brought about the adoption of a more "hands-on" policy for Canadian agriculture since 1972, but the primary reason is that Canadian governments have decided that the public at large must share with farmers the risks of agricultural production and future output expansion in an era that seems bound to be characterized by considerable instability in producers' returns and in their costs. An expanded public commitment to enhancing stability in farming has become the dominant theme of Canadian agricultural policy.

Numerous farm program changes have been made under the rubric of stabilization policy. Under the Western Prairie Grains Stabilization Act, grain producers in the Prairie provinces have been given a guarantee that their aggregate net cash flow from grain sales (gross receipts less current operating expenses) in any year will not be less than the previous five-year average. A revised Agricultural Stabilization Act has given producers of a wide range of crop and livestock products assurance that their unit returns will not be less than 90 percent of the previous five-year average market price adjusted for changes in current operating expenses. Administered prices for fluid milk and support prices for manufactured milk have been advanced to offset fully any cost increases. In addition, the federal government has felt the need to insulate the Canadian market for livestock products from the violent and unfettered fluctuations in supplies and prices in the larger and disordered U.S. livestock sector. This was accomplished by imposing temporary border controls on trade with that country in cattle and beef and imposing import regulations on poultry meats and eggs as part of national supply management schemes for the latter commodities.[1] In all, about 90 percent of Canadian farm output is now covered by federal stabilization measures.

---

1   For an extensive analysis of these developments, see Canadian-American Committee, "The Beef Story: Evolution of a Canada-U.S. Trade Issue" and "Canada's Import Quotas on Eggs: A Case of Domestic Goals Versus International Relations," *Backgrounder No. 1,* May 1975 and *Backgrounder No. 2,* September 1975 (Washington, D.C. and Montreal: National Planning Association and C.D. Howe Research Institute).

Federal government initiatives in the area of agricultural stabilization have been supplemented by programs developed by provincial authorities. Indeed, for some commodities, the center of gravity of agricultural stabilization policy is located in the provinces. The more modest of these schemes merely fill in gaps in the commodity coverage of the federal stabilization programs or provide somewhat higher price guarantees for particular products. However, some provinces have developed more ambitious "income assurance" programs designed to provide minimum prices for selected commodities that cover all costs of production (on efficient farms) and yield negotiated minimum returns to farmers' labor, management and capital investments. Additionally, provincial marketing boards have tightened their control of supplies or prices for some commodities, and national regulatory agencies have been created or are pending for eggs, turkeys and broilers.

It will be apparent that two quite different economic philosophies have recently molded the evolution of farm and food policies in the United States and Canada. In the United States, the combination of a robust faith in the ability of market forces to allocate resources and incomes in agriculture better than government decisions, confidence in the ability of commercial agriculture to accommodate substantial market instabilities, and a determination to avoid a return to close regulation of and heavy public expenditures on the industry has led to a marked reduction in the role of government in farming. By contrast, in Canada the conviction has taken root that the maintenance of an environment that will permit and encourage farmers to make the investments needed to meet growing domestic and world demands for agricultural products requires that market instabilities be reduced by the creation of a comprehensive set of public price and income stabilization policies for agriculture.

No judgment is offered here as to which approach will best ensure that the agricultural sectors of each country will be able to make their fullest contribution to national economic development and to meeting expanding world food needs. However, the effect on the bilateral relationship requires some comment.

First, neither country has grounds for complaint against the agricultural policies of the other unless they have a direct effect on conditions of competition in continental and offshore markets. Second, neither country has had a monopoly on the introduction of trade-distorting interventions during the past few years. Thus, it was the initial U.S. price controls on beef and export restrictions on grains, oilseeds and fertilizers that compelled Canadian authorities to introduce matching controls on trade in these products during the 1973–75 period. Conversely, Canada's action in placing eggs under import quotas in 1975 was the source of considerable bilateral friction. Third, Canada cannot be expected to permit the objectives of legitimate national agricultural stabilization programs to be jeopardized simply because U.S. policy makers and farm groups are willing to tolerate a markedly higher degree of farm level price and income stability.

There is no escaping the fact that the existence of more extensive arrangements for stabilizing agricultural markets on one side of the border than the other expands the scope for periodic bilateral trade disputes. How extensive this problem will become will depend on the objectives that are set and the techniques that are used in Canadian stabilization programs. The need for Canadian authorities to regulate bilateral trade flows will be minimized if agricultural stabilization programs go no further than the provision of stop-loss floor prices that maintain their relationship with competitively determined market prices, and if the deficiency payment is used as the primary instrument for implementing such minimum price guarantees. By contrast, the potential for conflict will be elevated if commodity-pricing objectives in Canada go beyond stabilization into the areas of full-cost recovery and income support, and if the latter objectives are secured primarily by the direct management of supplies permitted to reach the Canadian market.

At this juncture, it would appear that the greatest danger of restrictions on the access of U.S. producers to the Canadian market and of charges of unfair competitive practices attending Canadian exports to the United States and to third-country markets exists for trade in livestock and livestock products (other than milk) for which the long-standing and newly extended commitment of the federal and provincial governments in Canada to maintain stability has no parallel in the United States. Happily, in the all-important grain sector it would appear that there is little in the new program for enhancing stability in the Prairie grain industry that should give rise directly to bilateral trade difficulties, change competitive relations in third-country markets or hamper Canadian-American cooperation on a variety of matters relating to world grain production, pricing and trade.

NORTH AMERICAN AGRICULTURE IN INTERNATIONAL TRADE

Trade in agricultural products is one of the more important examples of economic interdependence between North America and the rest of the world. The region accounts for 20 percent of total world agricultural exports and for 10 percent of world agricultural imports.

Exports of both countries are dominated by the products in which comparative advantage is most firmly rooted: food and feed grains and oilseeds and their derivative products. Canada's farm product exports are quite narrowly based, with wheat accounting for half of all foreign sales. Almost three-quarters of the wheat produced in Canada is shipped abroad. Although the United States exports a much wider range of products, shipments of wheat, feed grains and oilseeds and their products recently have each constituted about one-fifth of that country's total agricultural exports. About one-third of total U.S. grain production is exported, and approximately 60 percent of the wheat and one-fifth to one-quarter of the feed grains produced each year are sold abroad. Foreign markets absorb about one-half of that country's soybean crop as beans, oil and meal.

Canada's share of world wheat exports recently has averaged around 18 percent. This is the only major traded agricultural commodity in which Canada has an important share of world trade. The United States has a much more dominant position in world grain markets, with its share of world wheat and feed grain exports standing at 45 and 55 percent, respectively. It is also the world's largest exporter of rice and soybeans, but trade in rice is small in relation to world production, while soybeans and their products are a part of a much larger market for fats, oils and protein meals.

The traditional markets of Western Europe and Japan and bilateral exchanges are still the most important commercial outlets for each country's exports of farm products. However, sales to the less-developed countries and to the socialist countries have developed rapidly, if erratically, in recent years. Exports to the OPEC countries have been particularly dynamic. The LDCs together absorbed about one-quarter of Canada's total agricultural exports in 1975 and 37 percent of U.S. farm exports in 1974–75. Whereas a high proportion of North America's agricultural exports to the LDCs used to be conducted on concessional terms, food aid accounted for only 5 and 4 percent of total U.S. and Canadian farm exports, respectively, in 1974.

The value of North America's agricultural imports has risen rapidly in recent years with the general rise in commodity prices. The United States is by far the leading supplier of Canada's imports of farm products, about 56 percent in 1974. A high proportion of U.S. agricultural imports originates in countries in South and Central America and the Pacific region: Brazil, Mexico, Australia, and the Philippines are the major suppliers. Approximately two-thirds and 80 percent of all agricultural imports of the United States and Canada, respectively, are products competitive with domestic production. However, imports of competitive agricultural products would be significantly higher were it not for the restrictive policies of both countries toward imports of manufactured dairy products, red meats and some other raw and processed food and fiber products.

Three features of North America's agricultural trade are of particular importance. First, the overall improvement in the income situation of U.S. and Canadian agriculture that has occurred in the 1970s is directly attributable to the rapid advance in the value of agricultural exports (see Table 1) and more particularly to the growth in the value of foreign sales of grains and oilseeds. Second, there has been a major shift in the structure of the trade of the United States and Canada, whereby a large positive balance on agricultural trade now offsets a part of the weakening trend in the balance of nonagricultural merchandise trade

TABLE 1

CHANGES IN NORTH AMERICAN AGRICULTURE, 1951-74

Item	Units	United States				Canada			
		1951	1961	1971	1974	1951	1961	1971	1974
Number of census farms	'000	5,430	3,825	2,909	2,830	623	487	366	320
Agricultural employment—	'000	9,546	6,219	4,436	4,313	939	681	510	473
As % of total employment	%	15.7	8.1	4.6	4.4	18.4	11.2	6.3	5.2
Average size of farm	acres	215	305	377	384	279	359	463	470
Aggregate net farm income	$ billion	15.2	12.0	14.4	27.7	1.9	0.9	1.7	3.8
Average net farm income per farm operation	$	2,739	3,040	4,950	9,211	3,100	1,893	4,590	11,820
Index of purchased inputs	1967=100	72	84	106	109	n.a.	85	105	119
Total capital employed	$ billion	132.5	204.8	317.5	475.9	9.7	13.1	23.9	36.1
Farmers' equity	$ billion	120.1	178.6	263.1	401.8	n.a.	11.3	19.2	29.4
Total farm debt	$ billion	12.4	26.2	54.4	74.1	n.a.	1.8	4.7	6.7
Capital per farm	$ '000	17.2	43.9	94.7	150.0	15.5	27.4	65.4	110.3
Capital per man	$ '000	9.4	22.2	55.2	90.3	10.3	19.1	47.0	76.1
Productivity index	1967=100	74	94	110	104	n.a.	81	115	102
Output per man hour	1967=100								
Farm		37	80	130	133	54	69	141	144
Nonfarm		65	70	110	110	63	85	115	101
Consumer food expenditures in relation to—									
Disposable income	%	n.a.	19.8	15.7	16.8	22.9	18.6	15.6	16.6
Expt. on goods and services	%	—	—	—	—	—	—	—	—
Value of agricultural exports	$ billion	4.0	4.9	7.8	21.3	1.0	1.2	2.0	3.8

Sources: *Agricultural Statistics* (Washington, D.C.: USDA), various years; *Selected Agricultural Statistics for Canada* (Ottawa: Agriculture Canada), various years.

(see Table 2) and deficits in other sections of each country's external accounts. Third, there has been a structural change in world agricultural trade, with a secular tendency for other regions to become dependent to an increasing extent on North America as the source of their grain imports (see Table 3).

Four profoundly important issues flow from these features of North America's agricultural trade. First, the future ability to sustain the recently experienced higher levels of aggregate farm income without a return to government support programs is crucially dependent on the continuation of a high level of foreign demand for North America's exports of grains and oilseeds. By the same token, the crop sector of the industry has never been more vulnerable to downward fluctuations in world import demand. It follows that foreign economic policies designed to ensure steady growth in the purchasing power of overseas buyers and to secure improved access to foreign markets are vitally important to the economic well-being of North American agriculture as a whole. However, the situation is complex. Trade expansion and liberalization would benefit producers of grains and oilseeds, while producers of some already noncompetitive livestock products would be doubtly disadvantaged by increased feed costs and intensified competition in their product markets. Hence, trade liberalization would tend to redistribute income within the agricultural industries of both Canada and the United States.

Second, the future performance of the North American economy is directly linked, via the balance of payments, to success in exporting agricultural products. More especially, the ability to finance the heavy cost of future imports of energy while maintaining an overall balance in external transactions and, more generally, the ability to reduce the real resource costs of North America's imports by supporting the international values of national currencies are dependent to a great extent on expanding overseas sales of those agricultural products in which North America has a clear international comparative advantage. This consideration also points up the importance of foreign agricultural trade policy initiatives designed to secure an expanding global demand for the products of North America's farms. Agricultural trade policies are examined in some detail in subsequent chapters.

Third, the reliance of a growing list of countries throughout the world on North America's ability and willingness to supply a proportion of their expanding grain and import requirements has raised the possibility that control over grain exports can be used systematically to achieve specific political objectives and enhanced diplomatic influence. The crudest concept of "food power" entails the deliberate withholding of grain supplies as a means of political coercion. For a variety of practical reasons this does not seem to be a plausible prospect. First, the concept is relevant only to the United States, for Canada has neither the global interests nor sufficient market influence to contemplate unilateral action, and it is difficult to identify circumstances in which Canada

TABLE 2

TRADE IN AGRICULTURAL PRODUCTS AND TOTAL MERCHANDISE TRADE, CANADA AND UNITED STATES, SELECTED YEARS

Year	Agricultural Trade			Total Merchandise Trade (Incl. Agriculture)			Agriculture Trade as a Percent of Total Merchandise Trade	
	Exports	Imports	Balance	Exports	Imports	Balance	Exports	Imports
CANADA				$ Can Million				
1951	1,020	711	309	3,914	4,085	−171	26.1	17.4
1961	1,193	813	380	5,755	5,769	−14	20.7	14.1
1970	1,685	1,283	402	16,458	13,952	2,506	10.2	9.2
1971	1,993	1,299	694	17,380	15,611	1,769	11.5	8.3
1972	2,135	1,538	597	19,500	18,736	764	10.9	8.2
1973	3,003	2,160	843	24,644	23,317	1,327	12.2	9.3
1974	3,813	2,828	985	31,292	31,578	−286	12.2	9.0
1975	3,916	2,891	1,025	32,096	34,547	−2,441	12.2	8.4
UNITED STATES				$ U.S. Million				
1951	3,448	4,557	−1,109	13,357	10,102	3,254	25.8	45.1
1961	4,963	3,794	1,169	20,853	15,308	5,546	23.8	19.1
1970	7,259	5,770	1,489	42,500	39,756	2,834	17.1	14.5
1971	7,693	5,823	1,870	43,492	45,516	−2,024	17.7	12.8
1972	9,401	6,467	2,934	48,876	55,282	−6,406	19.2	11.7
1973	17,680	8,419	9,261	70,246	69,024	1,222	25.2	12.2
1974	22,042	10,248	10,248	98,309	103,586	−5,277	22.4	10.4
1975	21,987	9,328	12,659	107,184	98,139	9,045	20.5	9.5

Source: *Survey of Current Business*, U.S. Department of Commerce, and *Selected Agricultural Statistics for Canada*, Agriculture Canada, various years.

TABLE 3

THE CHANGING PATTERN OF WORLD GRAIN TRADE,
BY REGION, SELECTED YEARS, 1934–76

Region	Annual Average: Million Metric Tons				
	1934–38[a]	1948–52[a]	1960/61[b]	1970/71[b]	1975/76[b]
North America	+ 5	+23	+39	+56	+95
Western Europe	−24	−22	−25	−30	−19
Australia & New Zealand	+ 3	+ 3	+ 6	+12	+11
Eastern Europe & USSR	+ 5	n.a.	0	+ 1	−36
Africa	+ 1	0	− 2	− 5	−15
Asia	+ 2	− 6	−17	−37	−46
Latin America	+ 9	+ 1	0	+ 4	+ 4

Notes: Inequality of imports and exports due to variations in reporting periods and different marketing years.
(+) indicates net exports, (−) indicates net imports
[a]Calendar years.
[b]Fiscal years.
Source: *U.S. Food and Agricultural Policy in the World Economy* (Washington, D.C.: U.S. Congress, Congressional Budget Office, April 26, 1976).

(and the other smaller exporters) would collaborate with the United States in a concerted program to withhold grain exports in pursuit of joint political objectives. Second, while the United States accounts for half of the world's grain exports, this represents less than 7 percent of the rest of the world's grain consumption. Most countries that conceivably might be the target of a coercive withholding operation would be inconvenienced only marginally by having to restrict grain consumption or find alternative sources of supply (including expansion of their own domestic production). Third, there is no consistent correlation between the dependence of particular countries on imports of U.S. grains and the desire of that country to influence their policies. And, even where such a coincidence exists, the leverage obtainable from the relationship at best would be minimal and highly variable over time.

Fourth, grain is such a fungible commodity that a selective embargo against a specific country would be difficult to make effective, while the alternative of a blanket restriction would hurt friendly states as well as hostile ones and bring particular harm to poor countries. Fifth, because the price elasticity of demand for U.S. grain exports is probably quite high in the medium term, restricting exports would entail some economic and political costs in the form of losses in foreign exchange and farm income. Sixth, there are important dangers to the open world economy that the United States has sought for so long to create and from which it derives great benefit in extending economic coercion as a feature of international economic relations. Finally, any attempt to achieve political ends by the deliberate creation or inadvertent intensification of hunger surely would attract a wave of moral revulsion at home and abroad.

Alternative concepts of "agri-power" have more appeal and greater plausibility. U.S. agriculture is a source of strength to the country's economy, and a strong domestic economy enhances the ability to influence world affairs. Beyond that, the greater goals of human progress, world political stability and harmonious relations between states demand that the productive capacity of U.S. agriculture be viewed not as another weapon in a national arsenal but as a unique global asset.

A fourth important issue flowing from North America's position in world agricultural trade is the scope of its productive potential. While the possibility of the predatory use of food power is probably a minor concern to policy makers in most countries, they are deeply concerned about two more immediate and interrelated issues—the physical ability of the United States and Canada to supply the rest of the world with a growing volume of exports of grains and oilseeds and the level and stability of the supply price of such exports.

The roots of their concern lie in the technological, economic and policy uncertainties that attend the future of North American agriculture. Important technological uncertainties are the size of the land area available to agriculture for cultivation, the rate of improvement in yields per acre and per animal and the possibility that weather will be less favorable to agricultural production in the future than in the past two decades.

A number of economic factors have disposed to the view that the long-run supply price of North American agricultural exports might tend to rise in real terms in the years ahead. On the demand side of the market, domestic consumption in the United States and Canada is sure to grow, and the goal of the foreign economic policies of both countries is to ensure an accelerated expansion in offshore demand for their farm product exports. On the supply side, factors that have moved the supply function of North American agriculture to the left are inflationary increases in most input costs, irreversible increases in the market prices of fuels and energy-related inputs, and the added costs of meeting higher environmental standards. The rate of improvement of productivity in agriculture—which counters these adverse cost developments—appears to have slowed down, and there are real doubts about whether agricultural productivity in North America can be accelerated quickly and easily. Also, the aggregate supply function of agriculture may have become less elastic with respect to price than in the past because of the decline in excess capacity in the crop sector of the industry. Future expansion of North American agriculture's output and exports will entail bidding labor and capital into farming in a high-wage and capital-short economy and making substantial new investments to increase the size of the land base. There are policy uncertainties faced by overseas countries concerned with the role that should be played by imports of agricultural products from North America in their food supply strategies. These include questions about whether domestic farm programs in the United States and Canada will encourage increased productivity and a stable upward trend in farm output, whether these countries will move at the international level to ensure availability of supplies and stability of price (by such means as participation in international arrangements on world grain stocks and prices), and whether they will permit unimpeded access to their supplies in periods of shortage.

From the few available studies of Canadian agriculture's production and export capacity, it would seem that the potential contribution of Canada to world food supplies in the future is easily exaggerated.[2] Contrary to the prevailing myth, the agricultural land base in Canada is limited. Only about 170 million acres are presently farmed (less than the total area of farm land in the Dakotas, Nebraska and Kansas), and there is not much additional land that could be brought into agriculture that would be suitable for sustained cropping. Three-quarters of the total agricultural area is in the Prairie provinces, lies above the 49 degree parallel and has a sparse and variable rainfall. The cropland base is unlikely to increase by the mid-1980s, and Prairie farmers are unlikely to change their summer fallowing practices. Of the 170 million acres now farmed, only 95 million acres are cropland, and summer fallow accounts for about one-quarter of the total. The area actually planted to all crops in the mid-1980s in Canada might be no more than 75 million acres. Domestic feed requirements will absorb the output from approximately 45 percent of this area, and additional acreage will be required for crops for domestic consumption and seed. Thus, it seems improbable that the combined supplies of wheat and barley available for export in 1985 could exceed 20 mmt., and beyond then they could fall.

The production and export capacity of U.S. agriculture has been subjected to more detailed analysis.[3] The results are more reassuring,

---

2   For instance, see OECD, *Study of Trends in World Supply and Demand of Major Agricultural Commodities* (Paris, November 1975), Chapter III; and Science Council of Canada, *Population, Technology and Resources,* Rep. No. 25 (Ottawa, July 1976), Chapters VI and VII.

3   For example, see G.E. Brandow, "American Agriculture's Capacity to Meet Future Demands," *American Journal of Agricultural Economics,* Vol. 56, No. 1 (December 1974), pp. 1093–1112; D. Droskin and E.O. Heady, *U.S. Agricultural Export Capabilities,* CARD Report 63 (Ames, Iowa, December 1975); A. S. Rojko "Estimating Future Demand: Alternative Grain Projections for 1985," *World Economic Conditions in Relation to Agricultural Trade,* WEC-10 (Washington, D.C.: USDA-ERS, June 1976); C.J. Yeh, L.G. Tweeten and L. Quance, *United States Agricultural Production Capacity: Preliminary Projection to 1985* (Washington, D.C.: USDA-ERS, November 1975), mimeographed.

since all the studies concur that U.S. production and exports of grains and oilseeds can be increased significantly by the mid-1980s over the higher levels that have prevailed in the 1973–76 period. Further, this can be accomplished with known technology and largely from the existing land base, though the area given over to crop production could be expanded substantially if needed. The studies differ in their detailed assumptions and results, but the following general picture emerges. Production of wheat could rise over the next decade to 70-80 mmt. from the 50-mmt. average produced in 1973–75, and most of the additional output would be available for export. Feed grain production could be expanded from the 1973–75 average of 175 mmt. to 260-270 mmt. by 1985, and export availability could rise from 40 mmt. to 60-70 mmt. Soybean output could be expanded by 50 percent, from around 40 mmt. in 1973–75 to 60 mmt. by 1985, and export availability could be twice the 15-mmt. average shipped abroad as beans and products in recent years.

The evidence on the supply price of this additional production and exportable supply is not conclusive. The supply functions for the U.S. grain and oilseeds sectors for an era characterized by rising energy and environmental costs and greater market fluctuations have not yet been determined. Clearly, a continuation of "favorable" price conditions would be required to call forth the upper levels of production and exports cited above, but precisely how high such prices would need to be in real terms is not known. However, it is worth noting the conclusion of one study that the above levels of production and exports would be reached comfortably (and, in the case of feed grains, exceeded greatly) if per bushel prices (in 1974 dollars) to farmers in 1985 were $4.00 for wheat, $3.00 for corn and $5.57 for soybeans.[4] Experience in 1975 and 1976 would suggest that U.S. and Canadian farmers would continue to plant "fence row to fence row" at somewhat lower prices. Certainly, there is no reason to believe that price levels would have to rise strongly above those experienced in the period 1973–74 to 1975–76 in order to call forth higher levels of grain and oilseed production and exports by 1985. In practice, effective world import demand for grains is expected to be lower than the level that would require U.S. agriculture to utilize fully its capacity for producing crops over the next decade.

---

4   Droskin and Heady, *U.S. Agricultural Export Capabilities.*

3

World Hunger and Food Insecurity

The deep anxiety that developed with the abrupt deterioration in the world food situation in 1972 seems recently to have abated. Contrary to fears and predictions, mass starvation in the less-developed countries was averted in 1972–73, and there has been some improvement in per capita food supplies in many (but not all) of the LDCs in the intervening years. However, the alarums of the past few years and the attention that has been devoted to world food issues have had lasting and positive results. The ignorance and complacency that prevailed about the world food situation have been dispelled. Instead, there is a sobering awareness that a majority of the world's people have but meager diets, that food insufficiency is a major cause of low productivity and premature death and that several hundred million people have a daily food intake that is barely sufficient to ensure their survival. Further, it has come to be appreciated that the immediate problem of improving the diets of the two billion people who now live in countries in which the overall food supply situation is unsatisfactory, and, beyond that, of ensuring an adequate supply of food for a rapidly expanding world population, are among the most urgent tasks facing mankind. Failure to remove the scourge of hunger from the present generation and to banish the specter of famine from generations yet to be born promises widespread human suffering and world political instability.

PROBLEMS AND SOLUTIONS

The dimensions of the problem are now well understood. For the world as a whole, increases in food production historically have kept half a step ahead of population growth, and per capita food supplies have been on an upward trend. However, "the world food problem" has two faces. In the developed countries, involuntary malnutrition is found in less than 3 percent of the population, and overeating is a major cause of disease and ill health among the remainder. The developing countries as a whole have made major advances in their food supply position in the postwar years. Despite all the difficulties they have faced, they have achieved an average rate of increase in indigenous food production as high as that in the developed countries, and they have succeeded in financing a rising volume of food imports from surplus regions. As a result, most people have eaten a little better each year, and the incidence of famine has diminished. Unfortunately, high rates of population increase in the LDCs have absorbed most of the gains from their rising agricultural productivity. Consequently, improvements in per capita food supplies have been small, diets remain inadequate and the incidence of nutrition-related diseases with a debilitating and deathly animus is lamentably high. The absolute number of hungry people in the less-developed countries is probably larger now than at any time in history. And, in years of crop failure or economic stress, hundreds of millions of people who normally live close to the margin of food sufficiency join the 400 to 500 millions who are estimated to be suffering from chronic protein-calorie malnutrition. In recent years, the rate of increase in food production has been slowing while population growth rates appear to have accelerated (see Table 4). Thus, the margin has narrowed.

Looking to the future, effective demand in the developing countries with market economies is expected to grow at an annual compound rate of 3.6 percent. This is higher than the rate of increase in food production that they achieved in the 1960s (2.8 percent per annum) or in the past five years (2.1 percent).

Unless the rate of growth of food production in the LDCs can be raised, there will be a rising gap between effective demand and indigenous supplies. This gap could average at least 85 mmt. of cereals by the mid-1980s, and in years of poor harvests it could be as high as 120 mmt. Technically, this quantity of grains would be well within the supply capacity of the developed country exporters, though this is a conclusion of no practical significance since there is but small prospect that the food deficit developing countries will have the foreign exchange available to sustain such a high level of food imports.

TABLE 4

AVERAGE ANNUAL RATES OF GROWTH
IN DEVELOPING MARKET ECONOMIES

	1952–62	1962–70	1970–75
Population	2.4%	2.6%	2.7%
Food production	3.1	2.8	2.1
Difference	+0.7	0.2	–0.5

Source: E.M. Ojala, "Trade Negotiations and World Food Security," address to the Trade Policy Research Centre, London, March 15, 1976.

This analysis of the plight of the developing countries led to the 1974 World Food Conference and to the articulation there of a comprehensive strategy for addressing world food problems. The strategy had four main elements: effecting an acceleration in the rate of growth of indigenous food output in the LDCs; enhancing world food security to avert dislocations in food consumption in poor crop years; arranging for adequate supplies of food aid to be available to needy countries; and changing world trading arrangements in ways that would permit the LDCs to meet more readily a part of their food requirements from world markets.

SHARED RESPONSIBILITY

At the World Food Conference, it was acknowledged that implementation of the strategy was a global responsibility and that all countries and groups of countries had a role to play. Along with a special moral responsibility because of their wealth, the developed countries had abundant self-interest in ensuring that widespread hunger did not jeopardize global economic and political stability. The nature of the developed countries' contribution and the manner in which the United States and Canada are meeting their responsibilities is the subject of the remainder of this chapter. But, before turning to that, it should be emphasized that the solution to the food problems of the poor countries is not to be found solely or even primarily in the policies adopted by the Western industrialized societies.

The food supply problems of the LDCs have been exacerbated materially by the actions of the OPEC countries. The rise in energy prices has dealt a devastating blow to the development prospects of the nonoil-producing developing countries, and lower rates of growth in their per capita incomes mean a prolongation of malnutrition. Further, the enormous current account payments deficits of the non-OPEC developing countries ($37 billion in 1975) and their growing indebtedness, attributable in large part to increases in world energy prices, have reduced their ability to import the agricultural inputs they need to expand domestic agricultural output or to purchase foodstuffs on world markets. Hence, the food problems of the majority of the developing countries would be alleviated significantly if the OPEC countries at best would cease to pick the pockets of the truly poor by initiating differential pricing of their oil, and, at a minimum, recycle a much larger part of their increased revenues from oil sales to the LDCs by expanding agricultural development assistance and by investing part of their surplus capital in the LDCs' agriculture-related industries.

Similarly, the developed socialist countries would make a real contribution to reducing hunger in the LDCs if their own agricultural sectors stumbled less frequently, if their domestic agricultural policies and food trade policies did not create instabilities in world food markets, and if they provided agricultural development assistance on a scale commensurate with their wealth.

However, the task of banishing hunger falls primarily on the authorities of the developing countries in which malnutrition presently exists and where the specter of future famine looms most ominously. The policy priorities lie in three areas. First, population growth rates must be reduced if per capita food supplies and the incomes with which to buy them are to improve. It is critically important that an early start be made on lowering fertility rates. UN median population projections indicate that zero population growth will not be approached in the developing countries until the middle of the next century, by which time their combined population will have increased from its present 2.8 billion to 10 billion. A delay of only 10 years in achieving zero population growth would raise it to 14 billion, meaning an increase of 40 percent in the number of people to be fed. Even by the year 2000, the difference between high and low population projections is greater than the present population of the developing world.

Second, since involuntary malnutrition is associated primarily with income inadequacy, policies affecting income distribution in the developing countries where malnutrition is a problem have a high priority. Inequalities in the distribution of income in developing countries are typically very wide, and there is both scope and need for measures that will enhance the incomes of the poorest sections of their societies.

Third, the developing countries must ensure that the pace at which the output of their agricultural sector increases is accelerated. The World Food Conference recommended that the annual average compound rate of increase of agricultural output be raised to a minimum of 3.6 percent, the rate at which effective demand was projected to expand. Thus, the LDCs were challenged to raise the annual rate of increase in their agricultural production by 40 percent or more and thereby meet most of the growth in their food demand from their own resources. The problems associated with attaining this objective are numerous, but there are few mysteries about what is needed for its accomplishment. Three requirements are preeminent: placing agricultural development higher on the list of national priorities; providing an expanded volume of resources for agricultural production and the development of rural infrastructure; and ensuring that rural cultivators have the incentive and the means to continually raise output by securing higher incomes from their rising productivity.

Time will reveal whether these requirements are met. This is a field in which scenarios to suit every taste may be portrayed. At worst, a combination of indifference and inaction with respect to both components of the food supply and demand equation could lead to the stork outpacing the plough and to human catastrophe on a horrendous scale. On the other hand, a determined effort to slow population growth rates and to utilize the genuinely enormous potential that is known to exist for expanding food production in the developing countries would ensure that their food situation need not be calamitous and indeed would continually improve throughout the remaining years of population explosion. In short, the technical and economic possibilities for providing improved diets for the rising populations of the developing countries exist and are well understood. If people continue to hover on the borderline of starvation, it will be because a key ingredient—political will—was deficient.

The contribution of the advanced societies should be regarded as supporting the efforts of the leaders of the developing countries. This contribution lies in four main areas: the provision of an expanded flow of development assistance; the supply of food aid; the enhancement of world food security; and the adaptation of world trading arrangements to permit the LDCs to make fuller use of international exchanges in their food supply strategies. The United States and Canada have central roles in each area by reason of their wealth, their capacity for leadership and their preeminent positions in world grain production and trade.

AGRICULTURAL DEVELOPMENT ASSISTANCE

Raising the rate at which food production expands in the developing countries to anything approaching the *minimum* target rate of 3.6 percent will require a greatly increased flow of financial resources into agricultural and rural development, most of which will have to come from the LDCs themselves. For many, this will require a wrenching reordering of their priorities for investment and current expenditures. However, the developed countries can help by directing a higher proportion of an expanding total level of transfers of financial resources and technology into agricultural and rural development. A commitment to do this was accepted at the World Food Conference. It was agreed that annual external transfers for agricultural and rural development should be raised from the surprisingly low figure of approximately $1.5 billion in 1972 to at least $5 billion per year (at 1972 prices) for the period 1975–80.

It is difficult to identify much development assistance with specific sectors. Though data compiled by the Consultative Group on Food Production and Investment indicate that the agreed redirection has occurred, inflation may have prevented the attainment of the target increase. Thus, the share of agricultural assistance in total aid flows rose from 12 percent in 1972 to about one-fifth in 1975. Combined bilateral and multilateral development assistance for agriculture rose to $2.5 billion in 1973 and $3.9 billion in 1974. Transfers of $5.0 billion were achieved in 1975 but only at current prices. Assuming that the lending of the newly created International Fund for Agricultural Development is additive, the flow of real resources for agricultural purposes will be augmented further in future years and the target should be reached.

The United States and Canada are clearly reordering their aid priorities. In 1975, U.S. bilateral assistance was concentrated on agriculture, rural development and population planning and the closely related fields of education, human resource development and health. In addition, policy guidelines governing U.S. food aid programs were revised to strengthen efforts to promote food production in poor countries. Similarly, following a major review of its international development programs, the Canadian government has decided that its assistance programs will be concentrated in the poorest countries and in five fields, of which food production and rural development are two. Both countries also have channeled an increased proportion of their development assistance through multilateral agencies such as the World Bank and supported the successful efforts of these institutions to raise the share of their lending for agricultural and food-related activities. In addition, it is notable that both the United States and Canada have recognized the particular importance of research in providing a solid basis for continuing increases in agricultural productivity. To this end, they have supported the network of international agricultural research centers, sought to strengthen local research capabilities in developing countries, and made attempts to refocus their national research priorities and resources on the challenge of world hunger. Though it takes time to effect changes in the volume and direction of development assistance, there is clear evidence that the governments of both countries have responded quickly to the imperatives of increasing food supplies in the developing world.

However, there is one North American resource that remains sadly underutilized. Food production in the developing countries would be improved if they had greater access to the capital, technological know-how and managerial skills of the firms and industries that constitute North America's agribusiness sector. Regretably, political hostility to foreign investment in many developing countries is a real obstacle to direct investment by firms engaged in the manufacture of agricultural production inputs and in food processing and distribution. The development of effective international codes relating to the treatment of transnational enterprises and of national insurance schemes to cover political risks, and progress in the UN agencies in devising equitable ways in which ownership, technology and operation can be "unpackaged," would bring the potential contribution of North America's agro-industries closer to realization. Meanwhile, since existing fertilizer, pesticide, machinery, and food processing plants in the LDCs appear to be plagued by operational bottlenecks and chronic undercapacity utilization, national agricultural assistance programs could well make fuller use of the technical and managerial skills of the personnel in North America's agro-industries.

One caveat should be mentioned about agricultural development assistance. The transfer of resources and agricultural technology from North America to the LDCs entails some sacrifice of the former's comparative advantage and potentially could result in a narrowing of foreign market opportunities for North America's farmers. Little is known about the dimensions of this effect. To date, the assumption seems to be that increased agricultural productivity in the LDCs would lead to faster economic growth rates and these in turn to higher personal incomes, improved diets and an increasing demand for North America's food exports. Though this may be true, the experience of 1975–76, when U.S. and Canadian producers of soybeans and rapeseed faced damaging competition in world markets from LDC exports of palm oil produced in part with financial and technical assistance provided by the two countries, should caution that the results of assistance programs may not be always or immediately benign. What is clear are that fewer political problems occur with development assistance directed toward increasing output of noncompetitive agricultural products and that trade policy initiatives to enhance the LDCs' ability to purchase part of their rising food requirements on world markets are an alternative to agricultural development assistance that may in general or in particular circumstances be much more appropriate.

FOOD AID

While the longer-term opportunities for expanding food production in the developing countries are great, in the near term there are a number of populous poor countries in which diets are inadequate, foreign exchange shortages are chronic and rapid improvements in agricultural productivity cannot be expected. For these countries, the provision of food on concessional terms is an immediate means of alleviating hunger. It was recommended at the World Food Conference that not less than 10 mmt. of food aid per year be provided to needy countries.

The United States and Canada have long been major providers of food aid and will remain so. Canada's food donations presently account for around one-quarter of its total development assistance, and it has announced that it will make 1 mmt. of grains available in each of the three years, 1975–77. Increased proportions will be routed through multilateral agencies and directed to the poorest of the developing countries. Historically, the United States has provided about one-third of all its economic aid to developing countries in the form of agricultural products made available as grants or sales on highly concessional terms. Unquestionably, the main motivation in the past was to reduce U.S. surpluses, and their distribution was not correlated closely with the incidence of hunger. But, under new legislation (PL 94-161), the emphasis has shifted to the use of concessional food shipments as develop-

ment aid, and a higher proportion is to be directed to the poorest countries. The United States also has taken the same important step as
Canada in entering into a multiyear commitment in physical terms. The
new law requires that 75 percent of food aid shipments must go to the
most seriously affected countries (MSAs) and be in grant form, and that
there be a minimum annual disbursement to the MSAs of 1.3 mmt. of
grains. U.S. food aid shipments were reduced drastically from earlier
levels in 1973–74 and 1974–75 but were restored to 6 mmt. in 1975–76,
representing two-thirds of all concessional supplies of cereals provided
to the LDCs in that year. Thus, the United States and Canada are
presently providing 70 percent of the 10 mmt. of food aid set as a
minimum target, and both countries have adapted their food aid programs in constructive directions.

Of course, the ancient controversies about the value and uses of food
aid are still unresolved. For instance, much remains to be done in the
administration of food aid programs to ensure that local food production
in recipient countries is stimulated rather than discouraged and that
supplies reach the most severely malnourished groups (including the
rural poor). In addition, two new issues require attention. The most
straightforward is the need to find an equitable formula for sharing the
costs among the developed countries of implementing the 10-mmt.
minimum annual commitment. One obvious method would be to raise
quotas under the Food Aid Convention of the existing International
Grains Arrangement from 4.2 to 10 mmt. A more difficult problem is to
ensure the availability of food aid and a reduction of its opportunity cost
in short crop years. This was not a matter of concern when the major
donors had burdensome stocks, but it has become a very real issue now
that excess stocks are no longer available. Recent experience has shown
that in conditions of shortage, food aid is an early casualty because of the
desire to grasp commercial sales opportunities and to hold down food
prices in the donor countries. It would appear that the ability to ensure
continuity in food aid shipments to recipient countries and to lower the
opportunity costs of such shipments to donor countries should be
counted among the benefits of a system of international grain reserves.

WORLD FOOD SECURITY

There are multiple ways of enhancing world food security. Accelerating
the rate of growth of total output, improving environmental control and
plant varieties to reduce the variance of yields, and improving global
information systems to better monitor the availability of supplies and
the development of demand are all measures with important roles.
However, the international debate on food security has centered on
proposals for multilateral action to rebuild and maintain world grain
reserves to levels that will provide insurance against future global
production shortfalls.

The world's reserves of wheat and feed grains were falling as a proportion of consumption prior to 1972 as the United States and Canada attempted to reduce the size of the surplus stocks they had accumulated under support programs. They were further and abruptly reduced in 1972–73 when world grain production fell, and poor crops in various parts of the world in 1974 and 1975 prevented their replenishment. Though some improvement is likely in 1976–77, the ratio of carry-over stocks to annual consumption will remain at levels that are considered by many to be too low (see Table 5).

Experience in the period 1972–76 has brought home the dangers of permitting the world's grain inventory to be drawn down too far. With

TABLE 5

WORLD YEAR-ENDING STOCKS OF GRAIN

	Wheat	Rice	Coarse Grains	Total Grains
1960/61–62/63				
Million metric tons	73.8	2.3	89.6	165.7
Percent of consumption	30.5%	1.4%	19.7%	19.3%
1969/70–71/72				
Million metric tons	84.1	17.5	71.9	173.5
Percent of consumption	25.1%	8.4%	12.2%	15.3%
1973/74				
Million metric tons	64.0	12.5	55.1	131.6
Percent of consumption	17.4%	5.8%	8.3%	10.6%
1974/75				
Million metric tons	61.5	12.2	48.5	122.2
Percent of consumption	17.1%	5.5%	7.7%	10.1%
1975/76				
Million metric tons	63.6	15.9	45.1	124.6
Percent of consumption	18.3%	6.8%	7.1%	10.2%
1976/77 (projected)				
Million metric tons	103.8	15.1	52.3	171.2
Percent of consumption	28.6%	6.4%	7.7%	13.4%

Sources: 1960/61 through 1975/76—*World Agricultural Situation*, WAS-10 (Washington, D.C.: USDA-ERS, July 1976).
Projections for 1976/77—*Foreign Agricultural Circular, Grains*, FG-29-76 (Washington, D.C.: USDA-FAS, December 15, 1976).

low inventories, poor current harvests result in multiple adverse consequences in poor and rich countries alike and do damage to the world economy and trading system. Poor countries are hard pressed to maintain already inadequate food consumption levels as available supplies are bid away by the rich. Concessional food aid supplies dry up at the moment they are most needed. The amount of food that can be purchased with financial assistance is reduced. The development plans of the LDCs are jeopardized as rising food import bills place strains on their balance of payments. And, in human terms, it is the worst fed people in the poorest countries who suffer the most. In developed countries, rising food prices fuel inflationary tendencies and redistribute incomes, and the livestock sector of agriculture is disrupted by increased feed costs. Interventions by governments in world trade flows are intensified, and importing regions are stimulated to follow more autarkic food policies.

For all these reasons, it has not been difficult for governments to agree that they should exercise collective providence to ensure that sufficient stocks of grains are on hand to mitigate the worst effects of periodic world grain crop shortfalls. The framework for this commitment was provided by the International Undertaking on World Food Security drafted by the FAO secretariat in 1973, endorsed at the World Food Conference in 1974 and accepted as a goal of the new international economic order at the seventh special session of the UN General Assembly in 1975. Agreement also was reached quickly that an international reserves scheme should consist of an internationally coordinated network of nationally owned and managed stocks. However, to date, efforts to translate this political commitment into a practical scheme with the force of a binding multilateral agreement have failed.

The most important reason for failure to reach an accord is disagreement on the purposes that a system of grain reserves should serve. Some governments, notably the United States, take the view that the primary purpose should be to ensure that total grain supplies (carry-in stocks plus current production) are sufficient to maintain global trend levels of consumption in poor crop years. On the other hand, the members of the European Community have insisted that the primary objective should be the stabilization of world grain prices by the concerted manipulation of stock levels. An agreement with the first purpose would have effects on world prices but need not contain specific price targets. Logically, decisions on its operation could be made primarily by reference to variations in supplies available. Conceptually, a reserves scheme with a consumption-maintenance objective could be created in the absence of an international commodity arrangement for grains. By contrast, a scheme having specific pricing objectives would constitute the core of an international commodity agreement for grains designed to hold prices between minimum and maximum levels, and decisions on stock accumulation and release would be triggered by movements in world market prices.

The only concrete proposal for the creation of a world grain reserve that has been formally advanced to date was presented by the United States in the International Wheat Council in September 1975. The plan is oriented firmly to the concept of creating sufficient national reserves in the aggregate to maintain trend consumption in the event of world grain production shortfalls. The proposal calls for the creation of a world contingency reserve of 30 mmt. of grain (25 mmt. of wheat and 5 mmt. of rice). This reserve would be over and above normal working stocks— estimated to be 35 mmt. of wheat and 15 mmt. of rice. Under this plan, the obligatory release of the contingency reserve would be triggered by shortfalls below trend in projected supplies. In the same way, obligations to acquire reserves up to the negotiated maximum would be governed by changes in stock levels and increases in world production above trend. It is claimed that a reserve of this size would be sufficient to cover 90 percent of anticipated world food grain production shortfalls from trend. Each participating country's share in the maximum stock would be determined by a formula reflecting the country's gross national product, its importance in world trade and the variability of its grain production. Under such a formula, the United States would hold 6.1 mmt. and Canada would hold 2.4 mmt. of food grains.

The U.S. proposal is modest in that it makes no provision for a reserve of feed grains and aims only at a 90 percent level of consumption protection. It therefore envisions the creation of a total grain reserve that would be a good deal smaller than others have suggested as desirable. For instance, the FAO secretariat has recommended that the world should attempt to hold 17-18 percent of its annual cereal consumption requirements in stock at the start of each crop season.[1] This translates into 220-230 mmt. of grains for the world as a whole or 150-160 mmt. excluding the USSR and China. The latter figure would be broken down between the different grains and between commercially held working stocks and the internationally agreed minimum contingency reserve as follows:

	Working Stocks (mmt.)	Contingency Reserve (mmt.)	Minimum Carry-in Stocks (mmt.)
Wheat	45	20	65
Rice	15	5	20
Coarse grains	50	20	70
Total	110	45	155

1   FAO, *World Food Security: Evaluation of World Cereals Stock Level,* CCP. GR. 75/9 (Rome, August 1975).

On the other hand, a grain reserve of 30 mmt. would be 50 percent higher than is required to meet the more limited objective of guarding against famine in South Asia—the world's most vulnerable area.[2] Further, it is extremely important to note that an international reserve of 25 mmt. of wheat accumulated and released by international agreement would be sufficiently large to hold world wheat prices within any range that those who favor a price-fixing international wheat agreement are likely to propose.[3] Hence, countries that favor a consumption-maintenance approach to food reserves and those that wish to see stocks used to stabilize prices between negotiated minimum and maximum levels are really not far apart in practical terms, though they continue to be separated by a yawning gulf of principle. Indeed, one might say that the dispute between the United States and the EC over whether price or quantity "triggers" should govern stock operations has gone altogether too far. This is because, in practice, the senior management committee charged with the task of monitoring international grain markets and recommending changes in the level of national stocks of necessity would have to consider both price and quantitative data, and their recommendations would have effects on both prices and supplies. That is to say, any operational scheme, whether consumption- or price-oriented, would need to use changes in both supplies and prices as presumptive indicators of the need for concerted action on international stock levels.

Such empirical studies as are available would suggest that the size of the reserve proposed by the United States is uneconomically large insofar as *measurable* costs of a 30-mmt. world reserve would exceed *measurable* benefits. Further, since the existence and use of reserves transfers income from producers to consumers and from net grain exporters to importers, the United States and Canada and their grain producers would incur net losses of welfare and income if the U.S. proposal were accepted and implemented. Even if this were the whole story, these costs might be regarded as a legitimate impost on two rich countries that have made a political commitment to enhancing the security of the food supply of developing countries. However, the available empirical work is deficient insofar as analysts have not been able to quantify some of the benefits that would accrue to grain-exporting countries from the availability of more adequate grain inventories. These include more stable domestic food prices; avoidance of periodic dislocations in national

---

2   P.H. Trezise, *Rebuilding Grains Reserves: Toward an International System* (Washington, D.C.: The Brookings Institution, May 1976).

3   This conclusion is derived from two empirical studies that have examined the relation between world stock levels and prices and an informed judgment about the EC's pricing objectives in a revised International Wheat Agreement. See W.W. Cochrane and Y. Danin, *Reserve Stock Grain Models, The World and the United States, 1975–1985*, Tech. Bul. 305 (University of Minnesota, 1976); and S. Reutlinger, "A Simulation Model for Evaluating Worldwide Buffer Stocks of Wheat," *American Journal of Agricultural Economics*, Vol. 58, No. 1 (February 1976), pp. 1-12.

livestock industries; a reduction in the cost of meeting food obligations; and a higher level of foreign demand for North America's grain exports in the long term resulting from the ability of rich and poor importing countries to place a greater reliance on trade to satisfy part of their food needs. That is to say, the U.S. reserves scheme contains important, but not readily measured, benefits for that country and for Canada as well as for the less-developed countries—the plight of which led to its formulation. It is a happy coincidence when one can do better by doing good.

Canada is committed to the task of enhancing world food security, has endorsed the FAO understanding and has agreed to work toward a practical plan for ensuring that world grain inventories are maintained at more adequate levels. However, it has neither endorsed the U.S. proposal nor has it advanced constructive suggestions for changes or proposed alternatives. This appears to reflect a preference for waiting to see what emerges from the gladiatorial contest between the United States and the European Community over the fundamental objectives of an international grain reserves scheme. Canada's silence on the U.S. proposal also may be attributable to doubts about the workability of the U.S. plan and to the desire which Canada harbors, but which the United States has not shared, to see a stocking scheme made part of a wider international agreement for grains. Though this is not yet an area of open conflict between the two countries, it is not one where shared goals are evident.

TRADE POLICIES

Although there is no doubt that the LDCs will have to increase food output greatly if they are to meet their growing food needs, excessive emphasis on increasing their degree of self-sufficiency runs the risk of encouraging some of them to follow food supply strategies that are inappropriate to their resource endowments. Furthermore, the economic interests of the United States and Canada, and of the grain sectors of their agricultural industries, would be well served if the LDCs were enabled and encouraged to make even greater use of external sources of food supplies. The LDCs as a group already import 10 percent of the grains they consume (34-mmt. average net imports in 1973–75), and they represent the most dynamic future outlet for North America's exportable grain surpluses. However, the degree to which they are presently able to use the trade option in their food supply strategies is constrained by the trade policies of the developed countries, and the willingness of the LDCs to rely on trade is further impaired by the risks associated with that course.

On the first aspect, the *ability* of the LDCs to earn a sufficient flow of foreign exchange to sustain a larger volume of food imports is constrained by the domestic and trade policies of the developed countries that distort international competition and frustrate the LDCs' comparative advantage in many lines of production. Thus, progress in liberalizing world trade for products of export interest to the LDCs has an important bearing on their future ability to expand their food imports. But, many LDCs consider that freer trade, while important, would not yield an increase in foreign exchange earnings sufficient to support the greater imports of foodstuffs and other products associated with their rising food needs and the accelerated economic development rates to which they aspire. Accordingly, they are seeking more far-reaching changes in world economic systems that would ensure that their per capita incomes and their earnings from trade were greatly expanded and made more stable. Although both aspects of trade reform—trade liberalization and more fundamental changes in international trading arrangements for the LDCs' commodity exports—are explored in the following chapter, it should be noted here that developments in this area will have an important influence on the rate of improvement of diets in the developing countries and on the size of the market they provide for North American agriculture.

Even if expansion of food imports were ostensibly the best use to which the LDCs could put any additional foreign exchange earnings they might secure from trade liberalization or from more radical changes in international trading systems, their *willingness* to extend their use of international food markets is affected materially by the economic and political vulnerabilities that accompany increased dependence on external food supply sources. The uncertainties and risks that LDC policy makers must take into their calculations when contemplating a trade-oriented food supply strategy include sharp variations in supply availability and prices due to output variations and the variable import demands of richer countries; the possibility of their being denied free access to supplies in periods of shortage; and the chance that they would be treated as residual customers to be served only when the bilateral commitments of the exporters had been fulfilled. That is, from the viewpoint of the LDCs, the factors that determine the level and stability of price and availability of supplies on world markets derive from the agricultural and food trade policies of the rich countries. Over these policies the LDCs have no control. Accordingly, initiatives in international trade policy designed to reduce the risks of relying on foreign food sources, and thereby to lower the discount factor the LDCs must presently apply to the trade option, are also important to the efficient use of the world's agricultural resources to alleviate hunger, and to the size of LDC purchases of foodstuffs from North America. Such measures include the creation of adequate food grain reserves; conclusion of an international commodity agreement for grains with price-stabilizing provisions; the negotiation of restraints on the ability of governments unilaterally to restrict access to their supplies; and progress in liberalizing agricultural trade (which would result in adjustments of prices to shortages being spread more widely instead of being concentrated in the international markets from which the LDCs must buy).

The interests of the United States and Canada in the above trade issues are direct and substantial because they require North America's leadership or concurrence, because they would entail changes in U.S. and Canadian domestic economic programs and foreign economic policies, because some of the proposed measures would benefit particular groups within North American agriculture and affect others adversely, and because not all the international trade policy changes involved are consistent with the objectives and interests of the two countries.

4

North American Agriculture and International Commodity Policy

One of the more dramatic changes that has occurred in world economic and political relationships in recent years is the success of the developing countries in shifting the subject of their poverty from the periphery of world affairs to the center. Among the reasons for their accomplishment are the use of their numerical preponderance and voting solidarity to ensure that their cause heads the agenda of all intergovernmental meetings; the success of OPEC as an exemplar of their expectations and supporter of their objectives; and the growing awareness of the advanced societies that economic growth and political stability in the world are dependent upon reaching an accommodation with the two-thirds of humanity that now constitute the "down-and-ins." The LDCs have become convinced that the world economic order is dominated by the developed countries and tailored to their interests and that nothing less than the creation of a new international economic order will permit them to escape from their poverty and give them the control over their destinies which they seek. Debate on the LDCs' demands was the subject of special sessions of the UN General Assembly in 1974 and 1975, of the fourth UN Conference on Trade and Development held in May 1976, and of the Conference on International Economic Cooperation (CIEC) held in Paris throughout 1976. The subject now permeates the work programs of virtually all the international institutions.

The concept of a "new international economic order" that the LDCs have proposed encompasses every facet of the relationship between the industrialized countries and the developing world: aid, trade, monetary arrangements, control over resources, access to capital and technology, the location of production activities, shared responsibility in decision making, and the structure and functions of the multilateral institutions. In each of these areas, the LDCs are demanding not marginal tinkering with existing arrangements but fundamental structural changes, the cumulative result of which would be to make their accelerated development a prime purpose of all international economic relationships. More particularly for our purposes here, the LDCs are

43

proposing a coherent and comprehensive strategy to deal with the special problems they face in trade in commodities.

THE UNCTAD INTEGRATED PROGRAM FOR COMMODITIES

As articulated by the Group of 77,[1] the so-called integrated program for commodities had seven principal elements: an expanding set of intergovernmental commodity agreements for an open-ended list of products; a common financing facility for those agreements with provisions for buffer stocks; index linking of the prices of LDC commodity exports to the prices of their imports; compensatory financial arrangements to guarantee the total value of their exports in real terms; a network of intergovernmental purchase and supply commitments; improved conditions of access to advanced country markets; and the deliberate transfer of primary processing activities from rich to poor countries.

These measures are intended to accomplish two broad ends. The first is to improve the performance of world commodity markets in economic terms. This entails reducing instabilities in prices so that market signals do a better job of allocating resources and incomes, allowing greater scope for comparative advantage to dictate the location of production and processing activities, and rationalizing marketing channels in order to eliminate wastes and excessive margins. The second goal is to improve the performance of world commodity systems in political terms, to be accomplished by adapting world commodity production, pricing and trade arrangements in ways that would ensure that there was a transfer of income and wealth from rich to poor countries.

The UNCTAD secretariat was charged with the task of translating the LDCs' demands into a set of practical proposals for action by the international community. A modified version of the secretariat's initial proposals was examined at UNCTAD IV (1976) and, with specific reservations being expressed by particular countries, was endorsed by consensus. It was agreed that preparatory meetings for international negotiations on individual products should be convened under UNCTAD auspices and that negotiating conferences should be completed by the end of 1978. A negotiating conference on the establishment of the proposed common fund is scheduled as this goes to press. Other parts of the program will be carried forward in the GATT multilateral trade negotiations, in CIEC and elsewhere.

The general objectives of the integrated program as now adopted are to create more stable conditions in world commodity production and trade, to reduce fluctuations in the LDCs' commodity export earnings and to ensure that their earnings from commodity exports improve in real terms at a pace adequate to support their accelerated development. These objectives are to be achieved by the introduction of a comprehen-

---

1   *Manila Declaration and Programme of Action*, TD/195 (Geneva: UNCTAD, February 12, 1975).

sive set of mutually supportive measures covering an extensive list of commodities. As originally formulated by the UNCTAD secretariat, the techniques were those listed above, and 17 specific commodities were singled out for priority attention.[2] The final resolution adopted at UNCTAD IV omitted an explicit reference to price indexation as a technique, and the list of commodities was changed by including vegetable oils and oilseeds and excluding wheat, rice and wool.

Commodities now on the priority list that are produced in North America are vegetable oilseeds and oils, sugar, meats, and cotton and cotton products. However, it is important to note that arrangements for grains remain of interest to the LDCs, though negotiations on them are currently being conducted outside UNCTAD. In addition, the list of candidate commodities is open-ended and will eventually be extended. Finally, many of the proposals for changes in trading arrangements will affect all commodities in which LDC and North American producers compete. Hence, it would be a grave mistake to believe that the integrated program affects North American agriculture only tangentially. On the contrary, if adopted, UNCTAD's multicommodity and multidimensional approach (both as to techniques and objectives) will encompass directly a large proportion of the output of North American agriculture.

The specific objectives of the LDCs in the negotiations that will be held in the coming months are to secure the application to commodity trade of six principal and mutually supportive measures.[3]

First, they seek the negotiation of a set of formal intergovernmental commodity agreements (ICAs) of indefinite duration. For 10 "core" commodities, the primary instrument of price management would be international buffer stocks.[4] The ICAs would provide for minimum and maximum prices for commodities moving in international trade. In principle, the negotiated price range for each commodity would bracket its long-run equilibrium price. However, reversal of an adverse trend in prices and assurance of minimum prices that would lead to an expansion of the LDCs' real export earnings are specific pricing objectives of LDC exporters and of the UNCTAD secretariat.[5]

---

2   The agricultural products were wheat, rice, meat, sugar, cotton, wool, bananas, cocoa, coffee, tea, rubber, hard fibers, and jute. The remainder were minerals.

3   Additional measures are also envisaged. These include demand expansion; improving the competitive position of natural products *vis-à-vis* synthetics; assistance with the extension of processing activities in developing countries and with export promotion; and changes in international market structures.

4   The 10 core commodities are sugar, cotton, cocoa, coffee, tea, hard fibers, jute, rubber, copper, and tin. The LDCs also anticipate that the negotiations now in progress in the International Wheat Council and the GATT will lead to an international arrangement for grains having stocking provisions.

5   *New Directions and New Structures for Trade and Development*, TD/183 (Geneva: UNCTAD, May 1976), p. 25.

Second, it is proposed that an international fund be established for those commodities for which buffer stocks are used to implement the pricing provisions of the ICAs. Resources for the common fund would be both subscribed and loaned, with part of the capital being paid up and part on call. Resources would be drawn from importing and exporting countries, and the OPEC countries and the international financial institutions also might participate. The primary function of the fund would be to act as a central banker for existing commodity agreements. Additionally, the fund would be used to give emergency support to prices of commodities for which ICAs had not been negotiated.

A third measure is the index linking of the prices of LDC commodity exports to the prices of their imports of manufactures. This is designed to prevent the erosion of the real value of commodity prices by inflation and changes in exchange rates. "Direct" indexation would entail changing the market prices of traded commodities to maintain the terms of trade at agreed base levels. "Indirect" indexation would require a system of international financial transfers (or deficiency payments) to compensate LDC exporters for shortfalls in the market prices of individual commodities below agreed reference levels.

Even where ICAs with minimum price provisions existed, the LDCs' commodity export earnings might fall below anticipated levels due to shortfalls in output or weakness in world demand. Accordingly, the fourth proposal is that ICAs be supplemented by a general arrangement for stabilizing the LDCs' receipts from their commodity exports by an expansion and liberalization of the existing International Monetary Fund (IMF) compensatory finance facility. The more ambitious demands of the Group of 77 would have the LDCs' commodity export earnings stabilized in real terms and around a rising trend.

Fifth, a wider role is seen for intergovernmental supply and purchase commitments in primary commodity trade. Their purpose would be to enhance market stability and predictability by a better matching of net export availabilities and net import requirements and to provide greater assurance of access to markets and to supplies for exporters and importers, respectively. Aggregate reciprocal purchase and supply obligations would be concluded multilaterally, but they would not specify the direction of trade, nor would they have the full character of a contractually binding obligation on individual governments.

Finally, the program calls for improved access to developed country markets for the raw and processed commodity exports of the developing countries. This would be accomplished by lowering tariff and nontariff barriers that presently impede the LDCs' foreign sales. The LDCs wish to see improved access accorded on a preferential basis. They place particular emphasis on the stimulus to local processing that would result from lowering the high effective rates of protection typically accorded processing industries in developed countries by the escalation of their tariff schedules with the degree of product fabrication.

IMPLICATIONS FOR NORTH AMERICAN AGRICULTURE

The demand of the LDCs for a new international economic order is a profoundly important matter for North American agriculture for three reasons. First, measures that will accelerate economic growth in the developing countries and expand their capacity to purchase foodstuffs from Western sources will benefit North American agricultural producers. Second, North American producers will be affected directly by changes in trading arrangements adopted for commodities produced in both the LDCs and North America. Third, the progressive establishment of an international regulatory regime for commodities in which government decisions grow in importance relative to market forces will effect fundamental changes in the international economic environment in which North American agriculture will have to function. The manifold ways in which the UNCTAD integrated program for commodities will affect the agricultural industries of the United States and Canada are examined below.

International Commodity Agreements

Insofar as the primary purpose of ICAs was the stabilization of prices, supplies and investment incentives, they might be looked on with some favor. The price mechanism in unregulated markets does a remarkable job of one of its functions, the allocation of available supplies. It is less successful in allocating resources and incomes in the short and medium term. Commodity markets are characterized by cycles of overproduction and underproduction and arbitrary redistributions of income. Accordingly, there is a case for contemplating the negotiation of ICAs with stabilization objectives.

History attests that it will not be easy to negotiate durable commodity arrangements of the classical type. Only five have been introduced in the postwar years. Experience has shown that the technical and economic characteristics of many commodities preclude successful market management. Further, for the ICAs that have been negotiated, the dynamics of market conditions have exceeded the flexibility of their provisions, so that the resultant maldistribution of benefits and costs has created extreme fragility in the arrangements. For these reasons, it will be extremely difficult to implement the UNCTAD proposal that a network of ICAs be negotiated for a large number of commodities in a predetermined time frame.

However, the major difficulties with the proposal concern three fundamental policy issues. The first is the status of ICAs in international economic relations. To date, a central presumption of the world trading system has been that commodity market management by intergovernmental agreement was a departure from the norms of international commerce, to be undertaken only in exceptional circumstances. Furthermore, the sole legitimate purpose of such arrangements was to reduce the amplitude of fluctuations of prices around their trend. This is not what the LDCs are proposing. They are seeking nothing less than the creation of a permanent regulatory regime for a growing list of commodities and demanding that ICAs be used to change price trends so as to redistribute international income in their favor. Thus, acceptance of the LDCs' objectives would require a fundamental change in the Western world's views of the purpose of ICAs. Second, because developed countries are large producers and exporters of many primary products while most LDCs are net importers of internationally traded commodities, raising commodity prices would produce perverse redistributions of world income. Third, the buffer stock is a wholly inappropriate technique for effecting sustained income transfers, and, should the ICAs proposed by the UNCTAD secretariat be diverted from the stabilization objective and onto the course of changing the terms of trade, it is inevitable that global production control and market sharing would be required. In addition, multiple pricing systems would have to be introduced to offset adverse income effects on LDC commodity importers. A corollary of these developments would be that production, trade and pricing would become a matter of continuous and detailed political determination. At the extreme, every alteration in production levels, market shares and absolute and relative prices required by developments in technology, comparative advantage, consumer preferences, and changes in supply and demand would become a test of political strength, with the influence of market forces being tenuous and the capacity of markets for self-correction being excluded.

The issues involved in the proposal to create an extensive set of ICAs transcend the particular effects that such a development would have on North American agriculture. In that specific context, however, it is not apparent that the immediate and long-term interests of the agricultural industries of the United States and Canada would be furthered if wheat, rice, sugar, cotton, and oilseeds (and other products that might be added to the LDCs' open-ended candidate list) were drawn into a generalized regulatory regime for commodities. A case-by-case approach to stabilization-oriented ICAs can be supported, though with some skepticism about whether practical and durable arrangements can be devised. To concede that government decisions rather than market forces should be the dominant factor in commodity production, trade and pricing could pose a grave threat to the ability of North American agriculture to maximize its contribution to the U.S. and Canadian economies and to world agricultural production.

Common Funding

The LDCs and the UNCTAD secretariat regard the proposals for common funding of commodity-stocking arrangements as the second important pillar of the integrated program for commodities. It is argued that the availability of financial resources would catalyze the formation and facilitate the functioning of ICAs and that the aggregate of financial resources required under a joint funding arrangement would be lower than if each ICA were financed separately. Additionally, the LDCs attach great importance to the establishment of a new agency, potentially under developing country control, with broader powers to intervene in markets than those required merely to support buffer stocks.

The case for common funding is not compelling. It is doubtful if a shortage of financial resources has been a major obstacle to the conclusion of ICAs. It is improbable that (surplus) stocks would be an attractive financial investment outlet for the OPEC countries or any other group. More generally, it is impossible for governments of the Western countries to make a political commitment to establish a common fund in advance of an agreement that ICAs with stocking provisions are necessary and prior to testing the negotiability of mutually advantageous arrangements. In any event, the main danger to North American agriculture could lie in the proposal that the fund have an independent trading function. Two matters are cause for concern. First, one may doubt whether the expertise required for (multi) commodity trading would be available to the fund's managers. Second, the ambiguities surrounding the control of the fund's trading activities raise the possibility that there might be considerable potential for cross-commodity subsidization and for market disruption.

Indexation

The proposal of the LDCs that the prices of their exports be protected from erosion in real terms by indexation amounts to a demand for a system of internationally guaranteed prices for their commodity exports and for the continuous adjustment of these prices to offset any decline in the terms of trade between their exported commodities and their imports of manufactures.

There are numerous difficulties with this proposal. At the conceptual level, no direct inferences can safely be drawn between movements in the LDCs' net barter terms of trade and their economic well-being or their capacity to import. Technically, there are real difficulties in measuring changes in relative prices. General policy objections include the dangers that indexation would exacerbate inflation, impose additional burdens on commodity-importing poor countries and freeze international price relationships.

If indexation were implemented, the implications for North American agriculture would not be encouraging. Direct indexation—which implies adjusting the world market prices of traded commodities— would require continuous international supply management, market sharing and multiple-pricing systems. Indirect indexation, the suggested alternative technique, would in general be markedly less odious. Supply management, inflationary effects and perverse income transfers would be averted; it could be applied to a wide range of commodities; and greater selectivity in compensating poor country exporters would be possible. However, even with this technique, LDC exporters would have an advantage over North American suppliers of competitive products by being accorded an internationally guaranteed price.

Developed country opposition to indexation has been so strong that direct reference to it was left out of the version of the integrated commodity program that was adopted at UNCTAD IV. However, this was a tactic rather than a conversion. Indexation remains a central objective of the LDCs, and its retention was implied in the final resolution of the conference which called for the establishment of commodity-pricing arrangements that would take into account, *inter alia*, movements in prices of manufactured imports, inflation and exchange rate changes.[6] There is little doubt that the subject will resurface when detailed producer-consumer negotiations on the arrangements for specific commodities are joined.

Compensatory Finance

Western countries, including the governments of the United States and Canada, have favored the stabilization of the LDCs' earnings from their commodity exports by supplementary or compensatory financial arrangements. The existing IMF facility was extended and liberalized in late 1975, and the U.S. government proposed the creation of an even more generous "development insurance fund" at the seventh special session of the UN General Assembly in September 1975. This approach to helping the LDCs sustain their development plans and maintain their capacity to import foods and other goods has much to commend it. Compared with ICAs, compensatory financial arrangements attack the problems of instability in commodity earnings directly. There is more scope for selectivity in the distribution of benefits among developed country exporters, and market forces are permitted to play their resource allocation role.

From the viewpoint of North American agriculture, these arrangements have the demerits of underwriting the aggregate export earnings of LDC producers of competitive products. In addition, the competitive advantage of the LDCs would be enhanced to the degree that the grant element in compensatory payments was raised relative to the loan component and to the extent that entitlements for supplementation were determined on the basis of the "real" values of export receipts through indexation—both of which are goals of the LDCs. However, so long as the "norms" for the LDCs' commodity export earnings were projected on the basis of usual market shares and competitive prices, developed country producers probably would not experience a serious competitive disadvantage from the extension of compensatory financial arrangements.

---

6   *Report on the UNCTAD Conference at Its Fourth Session,* TD/217, Resolution 93 (IV) (Geneva: UNCTAD, July 12, 1976).

Multilateral Contracts

The proposal by the UNCTAD secretariat for the establishment of a network of intergovernmental contracts is difficult to evaluate, not least because this is an area of the integrated commodity plan where the objectives are most ambiguous. One component of the proposal is concerned with joint forecasting of net import requirements and net export availabilities and the exchange of information on production, stocks, prices, etc. This is a valuable exercise that long has been conducted in the Commodity Councils and the Intergovernmental Commodity Committees of FAO. Its extension to a wider range of commodities and participants would be welcome. However, it is clear that much more is proposed, including the acceptance by governments of importing and exporting countries or commitments to supply and to purchase specific volumes of products. The possibility of associations of exporting and importing countries overtly concerting their respective positions also is envisioned.

There are at least three difficulties with this concept. First, many governments of developed countries, including the United States and Canada, neither would wish to replace "arms-length" trading by private interests by direct government involvement in commerce nor have they the authority and the expertise required to do so. Second, it would seem inevitable that bargaining between associations of producing and consuming countries eventually would extend to institutionalized bilateral negotiation on prices as well as on traded volumes. To many observers, such a wholesale bureaucratization and politicization of world commodity trade would be a quantum jump in the wrong direction. Finally, even if the governments of the United States and Canada declined to participate in intergovernmental trading activities, U.S. and Canadian farmers still could be affected adversely by the concentration of the effects of global supply and demand fluctuations into the residual market not covered by contractual arrangements.

Improved Access

The reservations that the developed countries have about proposals for a generalized regulatory regime for commodities stem from their opposition to the violation of their fundamental trade principles and from their doubts about the practicability and worth of specific program components. This is not true of the demands of the LDCs for improved access to their markets. Here, all that is at stake is their economic interests.

If the developed countries take seriously the trade principles they espouse, a generous response to the demands of the LDCs for liberalization of trade in raw and processed agricultural products scarcely can be denied. Generally, the record of the Western countries on this score is not good. Tariff and nontariff barriers (including the provisions of national farm programs) remain high on LDC agricultural exports that are competitive with domestic output. Tariff escalation provides high rates of effective protection to many processing activities. Few competitive agricultural products have been included in the generalized system of preferences. As a result, the LDCs' comparative advantage in the production of some agricultural products is frustrated, and expansion of their export receipts and their vertical diversification into processing are correspondingly constrained.

The United States and Canada have a better record than Europe and Japan in this regard in that a high proportion of the LDCs' complementary agricultural exports already enter both countries duty-free; the United States has included a large number of agricultural and food products in its (belatedly introduced) generalized system of preferences; and agricultural policies for some competitive commodities (e.g., sugar and cotton) are now notably less trade distorting than previously. However, there is still great scope for further improving the LDCs' access to the Canadian and U.S. markets for competitive agricultural products. This could be accomplished by lowering the level and changing the structure of tariff schedules; expanding quotas; removing excise duties; reducing the protection accorded by specific farm programs; and by extending the agricultural product coverage, enlarging the quotas and relaxing the rules of origin of the generalized system of preferences.

Trade liberalization is an important element of the north-south dialogue on international commodity policy. It is the component that is most clearly coincident with the trade principles of the Western countries and, correspondingly, the measure to which they should be most ready to accede. There is no question that it would provide important benefits to the LDCs. It is appropriate, therefore, that high priority is being given to this topic in the GATT multilateral trade negotiations.

Trade liberalization for products of export interest to the LDCs would have diverse effects on North American agriculture. Commodity groups that look on the LDCs primarily as markets would gain from an expansion in the purchasing power of their customers. This is preeminently the situation for grain growers. By contrast, producers of agricultural products and suppliers of associated services for commodities that compete with similar products or close substitutes originating in the LDCs would experience intensified competition in the North American market. This group includes producers and processors of many fruits and vegetables, beef, mutton and lamb, wool, tobacco, oilseeds, sugar, and cotton.

In addition, the terms under which the LDCs are given improved access to rich country markets are important in determining the extent of the intensified competition that North American producers of competitive products would face in domestic and foreign markets. For instance, the LDCs are asking that they be given preferential access. If this were accorded by other developed countries, North American agricultural exports would face intensified competition in overseas markets. Conversely, if the United States and Canada sought to provide

LDC exporters with expanded market opportunities by lowering their most favored nation (MFN) tariffs, intensified competition in the domestic market would result from increased supplies from both LDC and developed country sources. There are other important aspects of the terms of liberalization. Thus, the LDCs are requesting that they be granted bound margins of preference, be permitted to use export subsidies, have their exports exempted from the application of safeguard measures, and be permitted to meet lower standards in the areas of health and sanitary regulations. Most of these requests seem too extravagant and will be resisted.

CONCLUDING OBSERVATIONS

A moment in history has been reached when the demands of the LDCs for a reordering of the world's economic arrangements can no longer be met with torpid inaction by the developed countries. The notion of a "new international economic order" has given conceptual coherence to the nature and scope of the changes the LDCs seek, and the "integrated program for commodities" has drawn together for simultaneous consideration the measures that the LDCs perceive as constituting a coherent global strategy for commodities. The outcome of the debate that is now under way between rich and poor countries on the nature of their economic relations, including appropriate arrangements for world commodity production and trade, is decisively important for the living standards of many hundreds of millions of people and for the achievement of amity in political relations between developed and developing nations during the remainder of this century.

Ostensibly, the scope for agreement between rich and poor countries has widened in recent years. The developed countries have recognized that the accelerated development of the LDCs is not "an optional extra" in the functioning of the world's economic systems but a moral, political and economic imperative. Equally, they have a clearer perception of their interest in dissuading the LDCs from "rocking the planetary boat" by attempting to interrupt supplies of essential raw materials or by blocking progress in a variety of forums addressing important global policy issues. Also, the developed countries well understand that the stable growth of their own economies requires free access to assured supplies of raw materials and an enhanced stability in the LDCs' ability to purchase Western exports. For their part, leaders of the less radical LDCs have moved away from ideological posturing and the presentation of impossible demands for implausible changes in the world economic order. Their concern now is to secure a more equitable share of a growing world product while avoiding damaging an admittedly imperfect, but also dangerously fragile, world economy. Both groups appreciate that if international economic relations are treated as a zero-sum game, the result will be a negative-sum outcome in which all will lose, and none more so than the poor.

There is much to applaud in the proposals that the UNCTAD secretariat has made for a coherent global commodity policy. The emphasis placed on cooperative actions that would improve the economic performance of commodity markets is particularly welcome. These measures include the more efficient use of world agricultural resources through the liberalization of trade and concerted measures to enhance stability. Cycles of underinvestment and overinvestment in agricultural products and the associated wild gyrations of supplies, prices, earnings, and expenditures are wasteful and disruptive. In the developed countries, commodity shortages feed price inflation, cause dislocations in consumption and stimulate investments in high-cost alternative sources of supply and substitutes. Excess production and low earnings disrupt the LDCs' development plans and impair their ability to purchase foreign goods, including badly needed foodstuffs. Hence, there are sound economic reasons for a cooperative search for means to liberalize commodity trade and to stabilize international commodity markets.

However, there should be no mistake about the gulf that still separates the developed and the developing countries. To date, the failure to find an accommodation in international commodity policy reflects a fundamental ideological division between the two groups on the functioning of the international trading system. Looked at from the perspective of the developed countries, three of the central assumptions they have held about the world trading system are challenged by the integrated program for commodities. First, it was assumed that, with temporary derogations and special assistance, the less-developed countries progressively would adopt the predominantly market-oriented system of international exchanges employed by the advanced countries and characterized by "arms-length" trading by private individuals responding to market signals. Second, there was an assumption that trade in commodities would fit, for the most part, into the same kind of international economic regime as trade in manufactured products. Selective concerted interventions by governments in commodity markets might be necessary on occasion, but these were to be regarded as aberrant and transitory, to be contemplated only when exceptional economic wastes could be demonstrated and implemented only when very favorable ratios between the benefits and costs of interventions were assured. A third assumption was that the international trading system was agnostic with regard to income distribution. Its central concern was efficiency in resource use and thereby the growth of world product, not its distribution. There was a presumption that if the distribution of world income resulting from competitive trade was politically unacceptable, redistribution should be effected by direct transfers and not by the manipulation of the terms of trade and market output and shares, since interventions of this nature are themselves prone to widen international inequalities in income and cause additional inefficiencies in world resource use.

Contrast these assumptions with the beliefs of the LDCs. First, they are convinced that economic wastes in unregulated and imperfect commodity markets are exceptionally large and that continuous intervention is warranted for this reason alone. Second, they believe that the supply, demand and structural characteristics of commodity markets are such that giving free rein to market forces will necessarily widen international income disparities. Third, and above all else, they hold the view that international economic relations should be concerned with equity as well as efficiency and that international commodity policy therefore should be directed toward effecting a redistribution of world wealth.

It will be apparent that the demands of the LDCs for the creation of a continuous, comprehensive, regulatory regime for commodities, with income redistribution its primary goal and with the levels and shares of production and the terms of trade established by political decision rather than by market forces, constitute a truly revolutionary challenge to the fundamental precepts of the world economic order. More particularly, it is at variance with the very basis on which the developed countries conduct their economic relations with each other and, beyond that, with their broad view of how the world trading system should evolve in the future.

Characteristically, the main burden of responding to the far-reaching demands of the LDCs has fallen on the United States. Canada has expressed sympathy for the LDCs' aspirations but has made few commitments and has been generally less explicit than the United States on its position on most of the LDCs' proposals. However, Canada's views are believed to be close to those of the United States on most of the key issues. The initial position of the United States was to maintain that the old economic order had served advanced and developing countries well; to deny that a new economic order was in the making; to stress that the primary concern must be with ensuring the growth of world output, rather than with its distribution; and to emphasize that adjustments in economic relations must confer mutual benefits on both rich and poor countries to be acceptable. Subsequently, however, the United States has advanced numerous specific proposals for changes in world economic systems that would favor the developing countries and particularly the poorest among them. All of its proposals are consistent with a liberal and a more just economic order, and many of them are coincident with the LDCs' aspirations; e.g., expanded aid, easier access to Western capital and technology, accelerated trade liberalization, more liberal compensatory finance arrangements, and a willingness to consider on a case-by-case basis the merits of commodity arrangements with short-run stabilization objectives. However, the United States has been resolute in its opposition to the elements that the LDCs regard as central—the use of commodity policy to transfer resources to the LDCs; agreement that intergovernmental commodity arrangements should be a permanent and widespread feature of world commodity systems; a prior commitment to the common funding of buffer stocks; indexation of commodity prices and export receipts; and the contrived redistribution of production and processing activities.

How the conflict of perception and purpose between the developed and the developing countries will evolve and which elements of their respective proposals will find an enduring place in future international commodity policies cannot be foretold at this time. Certainly, the answers will not be found exclusively in the teachings of Adam Smith and in the simple prescriptions of those with a simon-pure view of the virtues of competitive markets who reject all *dirigiste* elements in world commodity systems. Yet, neither will the answers be found in the writings of Karl Marx and in the proposals of those who seemingly would create a global command economy for world commodity production, pricing and trade. The answer lies somewhere in between, in some form of a "mixed" international economy that, like our national economies, is partly market-oriented and partly politically directed, and concerned with both efficiency and equity. At this juncture, the two are inseparably linked since the LDCs will not cooperate in the reform of the international economic system unless the problems of global equity are addressed, and their cooperation is needed for its effective functioning.

What is certain is that the agricultural sectors of the United States and Canada are caught at the center of the issues at this time of world change. As stated earlier, the industry as a whole has much to gain from the accelerated economic development of a group of countries that together already constitute a large commercial outlet for North America's farm product exports. It is also clear that some sectors of North American agriculture will face intensified competition as a result of the liberalization of agricultural trade that will constitute a component of the evolving trade relationships between the rich countries and the poor. But, it is equally certain that North American agriculture cannot escape the influence of global regulatory arrangements, the creation of which appears to dominate the drift of the times. The danger is that, in devising a new world order for commodities, the international community will contrive an inappropriate mix of market and political forces. Specifically, it is feared that by making a misguided response to the legitimate needs of the LDCs or by attempting to buy relief from continuous political harassment, the developed countries will agree to tilt the system so far toward the prescriptions of the LDCs that commodity production and trade will become subjected to political direction and management to a degree that impairs efficiency in global agricultural resource use and diminishes the potential of North American agriculture. The world is too poor for such a Faustian bargain.

5

Agriculture in the GATT Negotiations

Discussions on the future shape of trading arrangements among the advanced countries in temperate-zone agricultural products are centered in the multilateral trade negotiations (MTNs) currently in progress in Geneva under the General Agreement on Tariffs and Trade.

These negotiations are concerned with the traditional but elusive objective of securing freer conditions of international trade in farm products. However, three objectives are being sought in addition to freer trade. First, an attempt is being made to strengthen the code of commercial conduct to define more clearly what constitutes fair trade practice by amplifying the GATT rules on such matters as access to supplies, safeguard procedures and national subsidy policies. Second, the agricultural negotiations are concerned with devising measures that will enhance the stability of international agricultural markets. Trade liberalization alone has an important contribution to make to market stability, but the subjects of reserves and world price management are also involved. Third, the developed countries are committed to make a special effort to improve the situation of the LDCs with respect to their position in the international trading system both as importers and exporters of agricultural products.

The agricultural components of the negotiations are unusually complex. The trade arrangements that emerge must provide for both liberalization and stability. They must encompass commodities that are or may be in surplus or in short supply. In the former case, the task is to ensure that the burden of adjustment to unfavorable market situations is not forced on low-cost suppliers. In the event of shortages, the task is to ensure that exporters do not export food price inflation. Thus, the negotiations are concerned simultaneously with the conditions of access to markets and to supplies. At the end of the negotiations, national governments still must be able to deal with the social problems of their agriculturalists and with temporary instabilities in their food markets, but by methods that have minimal impact on the conditions of agricultural production and international trade. Relations between the nations

participating in the negotiations also are complicated. Individual countries enter the negotiations as both exporters and importers of farm
products, with export interests to promote for some products and import
practices to defend for others. And, while almost 100 countries are
involved in the MTNs, the negotiations on temperate-zone products
have a strong trilateral flavor, with issues between the United States,
the European Community and Japan dominating the agenda. In addition, trade in agricultural products is not isolated from other elements of
the negotiations. Issues being discussed in the working groups dealing
with tariffs, nontariff measures and safeguards are highly pertinent to
agricultural trade. What can be accomplished for agricultural products
is dependent on achievements in liberalizing trade in manufactures and
in improving general trade practices. The institutional framework also
is more complex than for previous negotiations, with discussions on such
important matters as grain reserves and pricing, the LDCs' trade interests and access to supplies being conducted in other forums as well as in
the GATT.

THE BROAD POSITIONS OF
MAJOR PARTICIPANTS

The prime objective of the major exporters of temperate-zone agricultural products is to improve their access to commercial import markets.
Essentially, this means reducing the level of protection accorded to
high-cost production by the trade regimes and domestic agricultural
programs of the importing regions and making world markets less
disorderly by curbing the use of export subsidies. Because the value of
agricultural exports from North America and Oceania to Western
Europe and Japan is many times greater than their imports of like
products from these areas, it is not possible to achieve mutual and
balanced advantage only by making reciprocal concessions within the
agricultural sector. Hence, improved access for farm products to the
major food-importing regions must be secured by offering tariff reductions on imports of manufactures and other concessions on trade practices.

Multilateral liberalization of trading arrangements for selected
farm products is perceived in the United States as an optimum trade
strategy in terms of the attainment of a mix of goals, including enhancing farm price and income stability; maximizing farm income; sustaining balance in external payments; and minimizing the need for government intervention in and budgetary expenditures on the agricultural
sector. Improved access to foreign markets for U.S. and Canadian farm
products will be crucial in the future if the recent gains in export
earnings, farm incomes and farm asset values should prove to be due to a
congruence of unusual and transitory events. Beyond its immediate
national economic interest, the United States has a broader purpose.

Agricultural trade is characterized by discriminatory regionalism, bilateralism and the widespread use of national policy interventions that distort production and trade patterns unfettered by the international rules of commercial practice embodied in the GATT. Hence, bringing this aberrant sector more surely within the framework of an open nondiscriminatory trading system, governed by rules of acceptable conduct, has implications for the continuing viability of the GATT and of a liberal international economic order.

The chief interest of the United States and Canada is in securing improved market access, primarily in Western Europe and Japan, for a very long list of commodities headed by food and feed grains. In the past, the United States has been opposed to an international commodity arrangement for grains with pricing provisions. In contrast, Canada seems favorably disposed, in principle, to an international grains arrangement. Relaxation of the import controls that the United States and Canada maintain on meats and dairy products is the major area in which North America faces demands from its trading partners.

Australia shares the interest of the United States and Canada in widened opportunities for its grain exports to Europe and Japan and, like Canada, also leans toward the conclusion of an international grains arrangement. Along with New Zealand, Australia also is seeking improved access for its exports of dairy products and meats to North America, Europe and Japan. The European Community shares the interest of Australia in opening up the North American market for dairy products and meats, partly because of its need to find outlets for its own exports, but also because larger imports by the United States and Canada would reduce the pressure of supplies from Oceania (and Argentina) on the European market.

Japan has been reticent about its negotiating stance on farm products. Japanese authorities are known to feel that the country's essentially unilateral surrender of its market for grains and soybeans requires few further concessions. Others take the view that Japan will have to pay for wider opportunities for its exports of manufactures and for assured access to supplies of foodstuffs by offering further improvements in access to its market for imports of a variety of agricultural products and processed foods. This will entail relaxation of the still formidably restrictive tariff and quota barriers that impede entry of foreign supplies of livestock products, fruits and vegetables and processed foods, and some simplification of the administrative procedures that make the Japanese market so difficult to penetrate. No major country is as dependent on foreign sources of foodstuffs to feed its people as Japan, and this dependence is destined to grow in the future regardless of the domestic farm programs it pursues. Consequently, it will be seeking arrangements that promise assurance of uninterrupted access to foreign supplies and stability of price, and it therefore can be expected to press for measures that limit the use of export controls and to favor multilateral commodity arrangements with price and stocking provisions and supply obligations for exporters.

The agricultural negotiations present extraordinary difficulties for the European Community. There are a number of groups within the Community (including most of the general farm organizations and commodity groups) that frankly would prefer omitting agricultural trade from the MTNs. This is clearly impossible, for if the overall negotiations are to succeed they must lead to trade advantages for the food exporters and greater market order and stability for all. Other groups within Europe have a more positive attitude toward the agricultural negotiations. Industrialists know that trade liberalization for agricultural and manufactured products is inextricably intertwined. Europe's consumers and taxpayers also are important constituents for more liberal agricultural trade arrangements. The United Kingdom and West Germany have long been dissatisfied with internal aspects of Europe's Common Agricultural Policy (CAP) and with the adverse effects on international economic and political relations of its external face. Policy makers in Europe, who are coming to view the CAP as a symmetrical policy that protects the interests of farmers in periods of surpluses and those of consumers in times of world food shortages, also judge that this dual role might be better fulfilled if the CAP were matched by international arrangements for farm products that provided more stability in world markets. Accordingly, the objective of the Community in the agricultural negotiations is to promote "an expansion of trade on stable markets in accordance with existing policies."

This carefully phrased statement implies two things. First, the essentials of the CAP (that there should be a single internal market, preference for Europe's farmers in that market and freedom to choose the instruments by which protection is accorded) are not negotiable. This does not mean that the external trade effects of the policy are non-negotiable. Second, in the Community's view, stabilization of world markets for the major temperate-zone commodities is a precondition to their liberalization. Here, the Community is saying that continuity of outlets for exports, and their possible expansion, must be accorded within the framework of agreements that provide continuity of supplies to importers and enhanced stability in world prices.

Consistent with this fundamental position, the Community has proposed that international commodity arrangements should be negotiated for the major grains, dairy products and sugar, with the provisions of such agreements being tailored to the needs of each product. The Community has indicated that the arrangements for grains should provide for minimum and maximum world prices implemented by the international coordination of national grains stock-holding policies, with purchase and supply obligations for importers and exporters respectively at the price ranges. No stocking provisions are envisioned in the international arrangement proposed for milk products. But, minimum world prices would be negotiated for the major dairy products, and importing and exporting participating countries would undertake to give preference in their purchases and sales to each other when prices were at the limits of the range.

It will be apparent that the position of each of the major developed countries on trade in temperate-zone agricultural products in the current MTNs is much the same as it was in the Kennedy Round of GATT negotiations. On the one hand, net exporters of particular commodities insist that the objective of the negotiations is liberalization of the conditions of trade in farm products and that this can be secured only by modifications of the protection accorded by national agricultural policies. On the other hand, no country is anxious to expose the producers of products that it imports to intensified international competition. Some of them, notably the EC, genuinely believe that the agricultural sector is fundamentally different from other sectors—by reason of its special social importance, the universality of government intervention and its unique propensity to instability—and that the priority task for international cooperation is to enhance market stability via formal intergovernmental commodity agreements.

There is no way of knowing in advance whether these wide gulfs of interest and philosophy among the developed countries—which have proved unbridgeable in all previous encounters within GATT—can be reconciled in the current negotiations. However, lines along which compromise might be sought and hopefully achieved can be suggested.

TECHNIQUES FOR IMPROVING ACCESS TO MARKETS

Providing the exporters with improved conditions of access to the markets of importing regions could be accomplished by a variety of routes. Conceptually, lowering tariffs and enlarging quotas is the simplest approach. Although tariffs are generally the least important measure used to bolster the incomes of agricultural producers, they are still widely employed, particularly for such commodity groups as fruits and vegetables and processed foods. The exchange of tariff concessions between Canada and the United States on a wide range of agricultural products was one of the more significant agricultural outcomes of the Kennedy Round negotiations. It can be anticipated that the mutual exchange of tariff concessions on agricultural products again will prove possible between all the major participants.

Similarly, quotas on agricultural imports are still very common, and the liberalization of some agricultural import quotas will be sought. In particular, this is an area in which the United States and Canada can expect to face requests for concessions on their import regimes for dairy products and red meats. At present, U.S. dairy import quotas limit the share of foreign supplies in domestic consumption to about 1 percent. Canada's current dairy policy aims also for a high degree of self-sufficiency in milk and milk products, with imports being tightly controlled and rarely providing as much as 5 percent of consumption. Similarly, imports of red meats into both countries are restricted by formal quotas and voluntary export restraint agreements with their principal suppliers. Under these arrangements, foreign supplies of beef account for only a fraction of continental requirements. Recently, some thought has been given in Canada to allowing dairy product imports to provide up to 10 percent of that country's consumption. This is suggestive of arrangements that might be examined in the MTNs—that is, U.S. and Canadian quotas and other restrictions on imports of dairy products and beef might be progressively relaxed according to a negotiated formula.

With respect to the United States and Canada seeking improved access to the EC, the above and several additional techniques might be feasible. Some of them would be equally applicable in dealing with Japan. The Community might agree to levy-free or reduced-levy quotas on imports of certain products. Possible candidates for such treatment are EC imports of rice and durum wheat, hard wheats entering the United Kingdom and Italy's feed grain imports. Of course, the problem with such arrangements is that they undermine elements of the CAP and might involve the EC's agricultural fund in additional expenditure. Nonetheless, they have been mooted by the countries concerned, and there are precedents for them. Limited arrangements of this nature might be negotiable.

A further possibility is that the Community might be willing to enter into agreements to purchase minimum volumes of specified commodities. The agreement entered into in 1964 by the United Kingdom with its major grain suppliers is suggestive of the type of arrangement that might be contemplated. Such agreements would need to be more than a "best endeavors" undertaking. To be worth anything in trade-negotiating terms, they would have to be treaty bound. The unhappy history of the U.K. Grains Agreement also demonstrates that the governments of the importing countries would have to be willing and able to manage the output of their domestic agricultural industries to give effect to such obligations. There also would have to be parallel undertakings by the EC on the use of export subsidies (and in the case of wheat on denaturing premiums) to prevent the import commitment being rendered worthless by an equivalent increase in exports or diversion of supplies to secondary uses. Commitments of this nature might be feasible for wheat and feed grains sold to Europe and Japan. They are also a possible method of assuring minimum access to the European and Japanese markets for shipments of beef from Oceania and Latin America, paralleling those now employed by the United States and Canada.

In the course of the Kennedy Round, one of the negotiating objectives of grain exporters was to secure such "access guarantees" from importers. As those negotiations proceeded, the idea was extended from the concept of a minimum purchase commitment to the much more far-reaching concept of negotiated maximum self-sufficiency ratios. This implies full-scale market sharing. The idea is not entirely foreign to North Americans. The arrangements that the United States and Canada have with their suppliers of red meats divide consumption between domestic and foreign supplies on an agreed basis, and the U.S. arrangement provides for imports to rise in step with consumption. Canada's tentative intentions with respect to manufactured dairy product imports also seem to be of this character. Such arrangements with the Community and Japan would have more appeal if they provided for the low-cost exporters' share of the market to rise over time according to an agreed schedule. They would have less to offer if they merely froze the present relative positions of domestic and foreign suppliers, and no attraction whatsoever if, as was proposed by the EC in the Kennedy Round, negotiated maximum self-sufficiency ratios were to apply to low-cost exporters too.

Another possible approach to improving access to the European market would be to secure in negotiations first the binding and second the reduction of its variable import levies on specific products. Although this would appear to violate one of the key features of the CAP—the variability of the levies imposed on imports of farm products—in practice some elements of the levies are designed to accord a fixed measure of protection to processing activities and should be just as open to negotiation as are tariffs. As an example, it might be possible to bind and reduce the levies on poultry products. The Community has agreed in the past to maximum levies on tobacco and some types of cheeses, so this option cannot be ruled out. Ceiling bindings on import levies would not interfere with CAP mechanisms in normal times, but they would force the Community to carry part of the burden of adjusting to conditions of surplus. European consumers would welcome arrangements that allowed internal prices to fall with world prices, and ceiling bindings would help the member governments to exert budgetary discipline over CAP expenditures.

The *montants de soutien* (MDS) concept offers a more comprehensive approach to reducing the level of protection accorded by domestic farm programs. This idea was first advanced by the EC in the agricultural negotiations of the Kennedy Round. The core of the concept is that the degree of protection given to each commodity in each country would be measured, bound in negotiations at a maximum level and subsequently negotiated downward. In its simplest form, the MDS would be the difference between the average price received by the producers of each commodity in each country and a set of "international reference prices." The latter might be a best estimate of world market equilibrium prices, or be the minimum prices in international commodity arrangements should these exist. Product-specific input subsidies might be included in the calculation of the MDS. The concept has the merits of expressing in a single measure the nominal protection accorded to each commodity in each country by the great diversity of instruments that governments use to raise the returns of their farmers, while leaving governments free to use whatever method of implementing the internationally bound maximum amount of support that best suited their circumstances. Negotiated reductions in MDS would allow an increasing proportion of the world's agricultural output to be produced under efficient conditions.

Paradoxically, the technique now finds more friends in Washington than in Brussels. And, as a practical matter, it must be agreed that the problems of measuring and policing MDS are by no means resolved. Nonetheless, the concept still is worth pursuing because it offers a comprehensive approach to bringing domestic farm programs under direct international influence, something which has not been possible heretofore.

STRENGTHENING THE GATT CODE

Access to Supplies

The international rules which apply to the use of export restrictions have become an important topic for the MTNs primarily because of the OPEC countries' restriction of oil supplies and the prospect that producers of other commodities might attempt to form exporters' cartels. The topic is also of immense importance to agriculture because controls on the exports of foodstuffs have been used extensively since 1973 by governments concerned with countering inflationary food price increases. Experience has shown that the use of export restrictions can snowball quickly as other countries are forced to protect supplies for their domestic consumers and regular overseas customers against demand deflected from countries initiating export controls.

The position of the United States and Canada on the question of access to supplies necessarily must be ambivalent. Considerations that make them reluctant to accept international restraints on their freedom of action to limit exports include the desire to retain the use of a counter-inflationary measure; the need to conserve supplies of energy and other nonrenewable resources in short supply; and the desire to limit exports of raw products in order to promote domestic value added. Canada has been explicit in expressing its lack of enthusiasm for an international code that would permit free access to its supplies of raw materials. However, there would be great dangers to the world economy if the use of export controls to export inflation, to exert political coercion and to effect economic extortion should become common practice. Furthermore, North American agriculture's long-term export opportunities would be diminished if the United States and Canada were to come to be regarded as unreliable suppliers of agricultural products. This would inevitably foster the desire for a still higher level of self-sufficiency in importing regions and the search for alternative sources of supply.

Accordingly, North American agriculture's interests would be well served if the U.S. and Canadian governments were to abjure the use of export restrictions and, beyond that, to exercise leadership in the MTNs to bring the use of export restrictions under international control. This might be accomplished by strengthening Article XI of the GATT or by reaching agreement on a supplementary code of conduct on export practices. Key provisions of such a code would include clarification of the conditions under which export controls might legitimately be used; acceptance of an obligation to give prior notification and hold consultations before new export controls were imposed; and acceptance of a requirement to demonstrate that serious injury to the economy would result if exports were not restricted. Additionally, the right to use export controls might be granted only for a limited period, and their operation might be placed under more effective surveillance by the members of the GATT. Because one can conceive of circumstances in which export controls were needed to protect against the predatory purchasing practices of a country that was not a party to an international commodity arrangement (e.g., to deny the USSR access to a grains-stocking arrangement in which it has declined to participate), the right to discriminate between buyers of restricted supplies would be necessary.

Had such an agreement existed since 1973, it is certain that the restrictions on exports of soybeans and grains by the United States and Canada, and the taxes placed on exports of grains by the EC, would have been far less damaging to trade confidence and less disruptive of world markets than they were.

Safeguards

Every country retains the sovereign right to protect a domestic industry from serious disruption by imports. This matter is of special importance to agriculture because all countries have been particularly ready to protect their farmers against price declines attributable to imports, and there has been widespread resort to import restrictions on agricultural products where market disruption was occurring or threatened. Indeed, next to textiles, agricultural products probably are the product area in which general safeguard action has been most frequently and most casually taken. Few countries have bothered to formally invoke GATT's provisions on safeguards, escape clauses or waivers when imposing controls on agricultural imports.

The lines along which the escape clause and safeguard provisions of the GATT might be strengthened in order to bring the capricious use of agricultural import controls under more effective international regulation might include the following elements. Voluntary export restraint arrangements would need explicit approval of the Contracting Parties and would have to be terminated if this approval were withheld. The definitions of circumstances in which market disruption could be claimed and of what constitutes "serious damage" would be clarified. Import restrictions imposed to prevent market disruption would be sanctioned only if the country could show proof of material injury prior to the imposition of controls. It would have to be demonstrated that imports were a major cause of market dislocation (and not just a "substantial" cause as is presently the requirement under the U.S. Trade Act of 1974). Import controls to safeguard a domestic industry would be sanctioned for a specific and limited period and would be conditional upon the availability of an adjustment assistance program for the industry if continued and intense import competition were anticipated. The right to discriminate in the application of import controls might be provided. And, all countries might be required to follow the practice of the United States of having applications for temporary protection examined in public hearings before an independent statutory body.

With so much inherent instability in agriculture, and with governments everywhere being so intimately involved in agricultural trade, farm groups in North America may be reluctant to have any constraints placed on the freedom of their national authorities to spring to their assistance when imports cause or threaten market disruption. However, trade is a two-way street, and, in the long term, all producers stand to gain from greater international discipline in the use of import controls.

Unfair Competition

This too is an area of commercial policy in which the agricultural interest is prominent. The general problem is that as governments extend the range of their involvement in economic affairs, the scope for their various domestic programs to affect the conditions of trade is widened. The effects of government programs on volumes and prices of goods offered for export or on the ability of domestic industries to compete with imports may be quite inadvertent (as, for instance, might be true of government assistance programs designed to promote regional dispersal of industrial activity) or implicit (as in the Community's actions in 1976 to require its feed compounders to incorporate surplus skim milk powder in animal feeds) or calculated (as with the payment of subsidies on farm product exports).

Concern with the effects of national economic policies on trade in farm products is long-standing. The whole postwar international debate on the trade distortions induced by national agricultural programs might be said to have been of this character. And, the long history of disputes over the use of export subsidies for farm products is but a particular instance of the problem.

The general issue has been elevated in importance and brought into sharper focus by the provisions of Title III of the U.S. Trade Act of 1974 which has extended the definition of measures that are considered to constitute unfair trade practices to include any bounty or grant to manufacture, production or export, and it reduced (after January 1, 1979) administrative discretion about whether countermeasures should be taken when unfair practices are determined to have been used. Countervailing action would be mandatory on products entering the United States with the assistance of improper aids, and retaliatory action would be taken against countries that were deemed to be competing unfairly with the United States in third markets. Under U.S. law, no injury test is required to trigger countervailing and retaliatory action.

An attempt will be made in the MTNs to bring more order and discipline both into the use of trade-distorting practices and into the responses of national governments to them. The subject is complex and its ramifications are wide. It would be romantic to expect that a final and comprehensive resolution will be possible in the current negotiations. But a start will be made.

A particular concern of the United States and Canada is to come to grips with the problem of direct export subsidies in agricultural trade. The subsidization by the EC of exports of dairy products, canned hams and beef to the United States, and subsidized competition in sales of wheat and flour, poultry and many other products in third markets, has been a source of mounting friction between the United States and the Community. Liberalization of access to the U.S. market for dairy products is predicated on the termination of the Community's export subsidies on such products. In the past, Canada also has lost overseas markets for grains and other products to subsidized sales from the EC and the United States. Looking down the road, this would seem to be a good time to prevent a recurrence of the competitive subsidization of exports of grains should the world grain supply situation ease.

There are a number of ways in which the subsidization of exports of farm products could be brought under control. For instance, observation by exporters of the minimum prices designated in commodity agreements would curb the use of export subsidies. The EC claims that this is one of the primary purposes and virtues of such agreements. However, the United States has proposed an alternative approach of much wider reach. Under this plan, an attempt would be made to establish a system of classification of *all* government aids, to distinguish between those measures that are acceptable and unacceptable in trade terms, and to link this classification to a code on the application of countervailing duties and other retaliatory measures. Under such an arrangement, national subsidies that had as their principal objective the encouragement of exports (or that caused substantial reduction in the production costs of products that compete with imports) would be "prohibited." Countervailing duties could be imposed on imports that benefited from such subsidies and retaliatory action taken against subsidization in third markets, and there would be no need to demonstrate injury by the country adopting such measures. Export subsidies on farm products would fall into this category; so too might mixing regulations. At the other end of the scale would be measures that had only tenuous connections with international trade, for example, grants for research, training or industry adjustment. Such aids would be "permitted" and would not constitute unfair competition and attract either countervailing duties or retaliatory action. Other types of national schemes for subsidizing production (or taxing consumption or intermediate production inputs), whether destined for export or domestic consumption, would be classified as "conditional" because countermeasures would be permitted only under certain conditions, the most important of which might be the demonstration of distortions to trade and the proof of injury. Canada appears to support the U.S. proposal insofar as it attempts to deal simultaneously with subsidies and the countervailing and retaliatory actions to which they give rise. The acceptance of an injury test by the United States in respect of "conditional" aids would be a major advance in general U.S.-Canada trade relations. However, Canada does not go all the way in supporting the U.S. plan, preferring a prior determination of injury and multilateral negotiations before countervailing or retaliatory action was taken even in cases in which the use of "prohibited" aids was detected.

Whether a new code on national subsidy policies and countervail and retaliatory measures can be negotiated in the MTNs remains to be seen. Governments will not readily agree to discuss their national economic policies and specific subsidy practices with foreigners, but such may be the price of interdependence in an open trading system. Certainly an approach along the lines sketched above contains the promise of throwing a net over the use of particular agricultural trade-distorting practices such as export subsidies, and of placing a more precise obligation on governments to discuss in an international setting aspects of their domestic farm programs that other countries consider to constitute unfair competition. Farmers in North America would both gain and lose if international agricultural trade relations were to develop along these lines. But, if they believe in a more open and orderly world trading system, their support should be given to measures that would subject unfair practices in agricultural production and trade to international constraint.

INTERNATIONAL GRAINS ARRANGEMENTS

Although all the areas discussed above are important to North American agriculture, the core of the MTNs for the United States and Canada must be the negotiations on grains. It is in the subgroup dealing with grains that the conflicts of interest and purpose between the major developed importing and exporting countries are most sharply drawn.

The EC has made it clear that it favors the negotiation of an international stabilization arrangement for grains. To date, the United

States has been equally emphatic in stating that it views such an arrangement with little enthusiasm and that it places higher priority on improved market access than on enhanced international market stability. Canada and Australia, the other major exporters, also seek improved access, but they are concerned that arrangements are devised to ensure a minimum price for their wheat exports, to avert the danger of competitive export subsidization should surpluses return, and to provide for multilateral sharing of the costs of providing concessional grain supplies to the LDCs. Hence, depending on its terms, Canada and Australia might favor a formal intergovernmental commodity agreement for wheat, and perhaps for all grains. Because Japan's main interest in grains is assurance of access to supplies and stability in price, it too might find advantages in a grains arrangement.

At present, international discussions on a grains arrangement are only at the stage of exploratory diplomacy. Their center of gravity is more in the International Wheat Council in London than in the MTNs in Geneva, though they must finally move to the latter venue. No comprehensive proposal on the provisions of a possible agreement has been tabled by any party, still less specific proposals on its crucial elements. Consequently, at this time, it is possible to discuss the subject only in broad terms.

The Community's stated objective is to secure the negotiation of a comprehensive grains arrangement with provisions for maintaining prices within negotiated ranges between which each of the major food and feed grains would move in international trade. Prices would be held within the ranges by the acquisition and release of nationally owned grain stocks of agreed size and distribution and according to a negotiated formula. In the aggregate, the national stocks would constitute a world buffer stock, and they would be the major instrument for managing prices in the commercial market. The total quantity of stocks held by participating countries would be sufficient to keep international prices within the range, under most circumstances. However, exporters would have an additional obligation to supply at the maximum price the quantities normally purchased by the importers if prices broke through the ceiling, and importers would have an obligation to obtain their commercial requirements from exporter-participants if world market prices reached the floor. An international agreement on the provision of food aid, on an agreed shared-cost basis, might be part of the agreement, or an adjunct to it. Potentially, there are advantages and disadvantages for both importers and exporters in such an arrangement.

For the net importers, the overwhelming advantage lies in the supply and price assurance that an international agreement would provide. This would derive from the availability of stocks for use in short crop years and the ability to purchase their import requirements at a maximum price. Additionally, a minimum price provision that assured acceptable returns to producers in exporting countries would encourage sustained world grain production. Immediately obvious disadvantages for the importers in the Community's general plan, as it now stands, include their inability to buy supplies in the short term at bargain basement prices on severely depressed markets, and the obligation they would incur to bear a part of the costs of stocking operations that previously have fallen only on the exporters.

Benefits to the exporters include some degree of assurance against the collapse of world grain prices and export earnings; the avoidance of export subsidy wars if floor prices were respected; relief from part of the burden of holding stocks in weak market conditions; and the possibility that some constraints would be placed on the price support and production policies of the importing countries if prices were falling and their stocks were mounting. It is possible, too, that the importers might be prepared to accept a lower level of self-sufficiency in the long term if world market conditions were more stable and predictable. Furthermore, it is not inconceivable that, given the assurance of supply afforded by stocks, importers would not need to insist on a supply obligation at the price ceiling and that they could be persuaded to accept a minimum purchase obligation at the floor. The feasibility of these latter two features can only be determined in negotiations, for they are not as yet part of the Community's plan. Tangible disadvantages to the exporters include the loss of extraordinary export earnings in short crop years to the degree that importers' requirements had to be provided at the ceiling price, and, conceivably, some reduction in trend earnings due to the continuous presence of larger average world grain inventories. However, the gravest weakness from the exporters' point of view is that the outline proposal makes no explicit provision for improving access to the importers' markets. Better access might result from some blunting of the importers' impulse to increase their degree of self-sufficiency and from the pressure on their domestic grain support programs of the costs of the stocking obligations they would assume. But, these indirect means of improving access are highly speculative. They are not the assured and measurable results that are meaningful in trade-negotiating terms.

No categorical judgment can be offered at this point on whether North America's agricultural interests would be served by participating in a grains agreement of the type which the EC appears to favor. All the really key issues—the target price ranges; the nature of supply and purchase obligations; the size and distribution of stock levels and their management rules; possible links between stabilization measures and provisions for improved access; and the feasibility and durability of arrangements pertaining to all these matters—have yet to be determined. At this time, the Community's overall plan is a concept rather than a proposal.

What can be said with some conviction is that from the viewpoint of U.S. authorities, the fundamental yardstick of success in the negotiations on grains in the MTNs is whether it is easier for low-cost suppliers to sell increased quantities of products into importing regions after the negotiations are completed than it was before they commenced. As they see it, enhanced stability of world grain markets is a useful secondary benefit that might emerge from the MTNs. It is not a substitute for reduction in the protection accorded to high-cost producers in importing regions. By contrast, Canadian authorities have given the impression that they lay at least as much store by arrangements that would stabilize world grain markets as by those that would expand them. If true, this is an important difference in approach to the grains negotiations between the two countries.

It is possible, of course, that the current exploratory negotiations between the net grain importers and the net exporters will lead to the negotiation of arrangements for grains that will provide both improved access for low-cost producers and greater stability for all. However, there has been no indication to date that this is the importers' intent. On the contrary, there is widespread suspicion that their emphasis on stabilization is calculated to deflect attention away from the more nettlesome issue of liberalization.

Finally, it is worth remarking that liberalization of trade in grains alone would have a powerful stabilizing influence on world grain markets. Further, it may be possible for the United States and Canada to secure improved access for their grain shipments to importing regions by offering concessions elsewhere in the MTNs and without participating in a stabilization-oriented international commodity agreement for grains.

CONCLUDING OBSERVATIONS

At this stage, no one can anticipate what eventually will prove negotiable on agriculture in the MTNs. The elements of a bargain between the developed country participants exist in the reciprocal interest of agricultural exporters and importers in improving access to markets and access to supplies; in enhancing international food market stability; and in bringing agricultural trade-distorting policies under effective international rules. Strictly within agriculture, each major country has restrictive agricultural practices that it will be asked to relax, and these can be exchanged for a measure of trade liberalization in commodities in which it has an export interest. And, concessions on trade in agricultural products will be balanced by freer conditions for trade in manufactured goods and by acceptance of new codes and rules in various areas of commercial policy.

For the United States and Canada, there are great advantages in freer trade in farm products. This is an area of economic activity in which both countries have a distinct overall comparative advantage, and securing for agriculture the original promise of GATT would ensure that this potential was realized in the future. Not all sectors of North

American agriculture are internationally competitive. Grains, soybeans, citrus fruits, and poultry meats are commodity groups that would flourish under conditions of freer trade. The dairy and red meat sectors, fruits other than citrus, vegetables, peanuts, and cotton are the areas where freer access to the U.S. and Canadian markets would cause the most wrenching adjustments, though hardship would be reduced to the degree that trade liberalization for these products was multilateral rather than unilateral.

At present, the prognosis for freeing trade in farm products is not good. Agricultural protectionism has proved impervious to the repeated efforts that have been made in the postwar years to bring agricultural trade securely within the GATT. The determination of all countries to obtain improved access for their export of farm products in the current MTNs seems more resolute than ever before, but no country has abandoned its commitment to protect sectors of its domestic agricultural industry in which it is a net importer. And, given present market uncertainties, neither Europe nor Japan is anxious to diminish the degree of their self-sufficiency in food.

Therefore, the prospect must be faced that the accomplishments in the agricultural negotiations again will be small. Some fear that yet another disappointment of the expectations of the major exporters of farm products—and particularly those of the United States—would intensify their dissatisfaction with the working of the world trading system to the point where they would, at best, limit their offers on manufactures and, at worst, decline to complete the negotiations, adopt a more aggressive stance in commercial policy matters, and show a disinclination to cooperate in other areas of international economic policy.

Such an outcome would be a grievous disproportion and perverse. Trade in farm products is destined to be a growth business whatever the outcome of the MTNs. In the all-important grains sector, an expanding world need for North America's low-cost supplies is assured. Within the MTNs, there are important mutual gains in liberalizing trade in manufactures and in selective strengthening of the GATT code. Beyond the MTNs, there are many areas of global economic relations requiring cooperation between the advanced countries. It would be tragic if disappointments in the agricultural trade negotiations resulted in the United States and Canada adopting a curmudgeonly attitude on other economic policy issues and withdrawing their leadership in international economic affairs.

6

Trading with the Socialist Countries

Great uncertainty attaches to the durability and pace of development of trade in agricultural products between North America and China, the USSR and its East European partners. This is particularly true for the USSR. Further, this component of world agricultural trade presents weighty problems for its participants and for the operation of the world food system.

In recent years, China has been a regular importer of significant and reasonably stable quantities of Western grains, mainly wheat. Its requirements have been serviced readily, and it has provided a valuable export market outlet, particularly for Canada (the United States tends to be a residual supplier). But, in the main, China has chosen to limit its involvement in the world food system and avoid dependence on external supplies. From all accounts, it has had remarkable success in providing adequate diets for a huge population from a slender and agriculturally inhospitable land base. Knowledgeable observers are concerned that the prolonged isolation of the Chinese scientific community from technological developments in agriculture elsewhere in the world will constrain its ability to expand indigenous output if national policy calls for an improvement in the quality of diets. In this case, China might choose to extend its import demands. However, there is no clear sign that this is imminent, and it must be assumed that China's participation in world food markets will continue to be marginal and cause correspondingly few problems.

Dealing with the USSR is a much more complex and difficult matter. By any standards, its sporadic forays during the 1970s into Western grain markets have been a mixed blessing. Insofar as one component of Western political strategy is the process of "Gulliverization"—that is, the multiplication of trade and other links between Russia and the West so that a retreat from detente becomes progressively more costly—there may be important political advantages in encouraging the USSR to become more dependent on Western food supplies. And, if sustained, sales to this market will be a valuable and much needed economic outlet for North American agriculture. But, the large year-to-year variations in Russia's agricultural import demands have been a very destabilizing influence in world food markets.

Three questions attach to the future participation of the USSR in the world grain economy. Will it continue to be a net importer of food and feed grains and protein feeds and on what scale? How can the destabilizing effects of its variable import requirements be reduced? How can it be induced or compelled to cooperate in the common tasks of managing the world food system? No definitive answers to these questions can be provided, but some observations pertaining to them can be offered.

The first matter is highly speculative. Experience in Poland and other Eastern bloc countries would encourage the view that there would be considerable political risk in abandoning the commitment to increase the supply of livestock products available to Russian consumers. Equally, the weaknesses in Soviet agriculture—poor natural resource endowment; vulnerability to weather variation; constant loss of skilled workers to urban pursuits; distorted price relationships and inadequate production incentives; diseconomies of scale in large farming units; and sheer mismanagement in a highly centralized system of decision making—seem so deep-rooted that there is no early prospect of an acceleration of the trend of output increases or of reducing the between-year variability of supplies despite the continuation of huge national investments in the agricultural sector. If valid, these considerations would nurture the view that dependence on imports of Western grains would continue on a gradual upward trend, but with grain import demand varying from little or nothing in the occasional good crop year up to the limits of port handling facilities and foreign exchange availability in years of poor harvests. The ability to secure low-priced supplies of livestock products on world markets in some years would be a further factor affecting the level of import demand for North American grains.

The variability of Russia's grain import demand places a great strain on the world food system, especially when reserve stocks are slender. In years of large Russian purchases, food price increases fuel inflationary tendencies in the rest of the world, feed grain price increases dislocate Western livestock industries and poor food-deficit countries are on the crack end of the whip as food aid shipments dwindle and the cost of commercial imports soar. In years when the USSR's import demand is low, grain prices and grain producers' incomes tend to be weaker. In addition, the unpredictability of inter-year and intra-year Russian demand complicates the task of policy making for food and agriculture everywhere. Various approaches to mitigating these problems have been taken by U.S. authorities. These include attempts to secure prior and more accurate information from the Russians on their import needs; systems of monitoring the sales commitments of private grain traders; informal but effective export controls in the form of prior approval requirements and temporary embargoes on sales; and the bilateral intergovernmental agreement signed in the fall of 1975.[1] *En passant*, consideration has been given to the creation of a single selling desk agency for U.S. grain exports, akin to the Canadian Wheat Board, but this proposal found little support.

For the United States, the problems of dealing with the Russians have been reduced materially by the bilateral intergovernmental agreement. Russia has been politically legitimized domestically as a regular customer for U.S. grain exports. The predictability of total Russian purchases has been enhanced. The within-year scheduling of purchases has been stretched. The danger of a rerun of "the Great Grain Robbery" of 1972 has been averted. The ability to limit sales if supplies are short means that U.S. consumers face less chance of a general surge in food prices, and livestock producers are in less danger of a disruptive increase in feed grain costs. Regular foreign commercial customers have more assurance that their import requirements can be met.

On the other hand, looked at from a broader perspective than that of the U.S. interest, the agreement has not stabilized the USSR's position in the world grain economy. Its involvement could still range between being a net exporter of 5 mmt. of grains to a net importer of up to 30 mmt. Hence, the bilateral arrangement has added little to stability in the world grain economy as a whole.

Indeed, it is hard to escape the more general conclusion that, although the bilateral arrangement has conferred some benefits on the United States, it has made solutions to the wider tasks of managing the global food system more difficult to attain. The USSR is in an enviable position. Its variable import demands for grains will continue to cause far-reaching disturbances in the world food system and the world economy. It has guaranteed access to up to 12 mmt. of grains from the United States (8 mmt. of wheat and corn and additional quantities of other feed grains). But, it is still able to play the role of a "freeloader" in the world food economy by declining to participate in such common tasks as improving global food information systems, enhancing world food security and food price stability, and providing agricultural development assistance and food aid to poor countries. And, the USSR's nonparticipation in these tasks makes the prospect of their accomplishment less sure. For instance, having taken on an obligation to purchase grains, there is no reason to suppose that the USSR will now accept a further treaty-bound obligation to hold stocks under a multilateral agreement on grain reserves. At best, it is questionable whether a world reserves scheme could be effectively operated without the participation of a major trader and a prime source of the

---

1   The agreement commits the Soviet Union to purchase annually for 5 years (1976–81) a minimum of 6 mmt. of wheat and maize (in approximately equal proportions) and provides for the purchase of an additional 2 mmt. annually without intergovernmental consultation. Purchases over 8 mmt. require approval. The United States may block sales if national grain supplies are below 225 mmt.

instabilities the reserves are designed to attenuate. At worst, the very negotiability of a reserves arrangement may have been prejudiced. Nothing could better illustrate the limits of partial and bilateral approaches to the solution of problems with multilateral dimensions.

So far as Canada is concerned, the U.S.-USSR bilateral agreement could have unfortunate effects. The multi-year framework agreements that the Russians previously have sought with Canada (that were, of course, equally unhelpful to the solution of global food problems) may now seem less attractive to them. In years when harvests in Russia are good, it likely will meet its purchase obligation with the United States by buying feed grains and selling wheat. In this event, Canada would face the USSR as a competitor in world wheat markets, and it is a notably "softer" and less cooperative seller than is the United States. Conversely, if Russian grain purchases from the United States were limited in years of short crops in that country and in other parts of the world, there would be a danger that the USSR would attempt to empty the granaries of Canada (and Australia) and thereby jeopardize Canada's ability to meet the needs of its regular customers. Thus, the variability of the USSR's demand for Canadian wheat supplies could now be greater than it was previously.

Ad hoc accommodations to the problems of dealing with the USSR have been necessary, and they have had their successful features. However, at least two deep-rooted issues remain unresolved. First, the very method by which centrally planned economies conduct their economic transactions with market economies poses problems. There have been instances in agricultural trade in which the skillful exploitation of unequal access to information between monopsonistic procurement agencies and uncoordinated sellers has resulted in an asymmetrical distribution of the benefits of economic exchanges. Indeed, because central planners can consider the totality of the implications of their external transactions, while externalities and secondary effects are not perceived in advance (or cannot readily be internalized) in market economies, it may be that there is an inherent propensity for socialist countries to capture a disproportionate share of the gains from foreign commerce. Second, short-term solutions to the problems caused by unstable and unpredictable grain export demand can force international economic relations in agriculture in directions that lead away from preferred goals. Thus, while the political necessity and the economic advantages of the 1975 U.S.-USSR agreement on grains are well understood, it nonetheless extended bilateralism, discrimination and government intervention in international trans-actions. These are precisely the features of trade in agricultural products which U.S. commercial diplomacy has long sought to eliminate. On both counts, the conclusions might be drawn that the conduct of trade between groups of countries with radically different economic systems may require a multilateral code of trade rules that differs substantially from those embodied in the GATT and continuous political direction.

Whether changes in agricultural marketing institutions also are required is open to debate. Canada has derived some advantages from the fact that wheat marketing is controlled by a central agency, the Canadian Wheat Board. These advantages include the ability to deal head-to-head with import monopolies on the basis of near parity of information; to ensure that an appropriate level of stocks is always on hand; to enter into "multi-annual" framework agreements; and to systematically allocate supplies between the USSR and other claimants on national output. These advantages are not inherently available in the United States where grain marketing is in the hands of private firms. It would not appear, however, that there is a clear need for the United States to contemplate adoption of a centralized selling system now that its trade in grains has been brought under effective political regulation by fiat and by the bilateral agreement with the USSR. Moreover, many believe that if the board system were transposed, there would be significant economic costs in the forms of loss of efficiency in price formation, reduced merchandising agility and lower rates of innovation in grain marketing technologies and practices.

7

The Nexus of Farm, Food and Trade Policies

It will be apparent from earlier chapters that it is no longer appropriate to think of farm policy, foreign agricultural trade policy and food policy as distinct subjects and in that order, though it is only a slight exaggeration to state that this has been the practice in both the United States and Canada until very recently.

The North American agricultural and food system occupies a pivotal position in the attack on world hunger and as such is woven into the fabric of one of the most intractable problems facing human society. Agricultural and food issues figure prominently in the economic and political relations between North America and each of the politico-economic groupings of countries with which the United States and Canada must deal in an increasingly pluralistic world. The growing interdependence between North American agriculture and the world economy generates forces from without that dictate national priorities and mold national decisions concerning the domestic agricultural and food system. By the same token, the growing number of issues pertaining to agriculture and food on which U.S. and Canadian authorities must engage foreign governments provides new opportunities for advancing their national agricultural and other economic interests and for pursuing foreign policy objectives in the international arena.

At the same time, the agricultural and food system of each country is firmly embedded in the larger structure of their national economies. As a result, national economic policy concerns—economic growth, employment, internal price stability, and external payments balance—generate forces from within that dictate the objectives and the content of foreign trade policies for agricultural products. Thus, national policy makers need to give simultaneous consideration to the external effects of changes in national policies and to the domestic consequences of changes in the external environment.

In short, national economic policy making for the food and agricultural sector increasingly must serve the needs of a growing and stabilized national economy that is embedded in an interdependent world economy and an embryonic global polity.

These developments are recognized in the view that has found expression in both the United States and Canada that the needs of the time require that each country should develop "a comprehensive national food policy." This means that national policies for farming, food and foreign agricultural trade should be developed in concert and be synergistic. It also means that they should contribute to national economic progress and stability and to the pursuit of national economic and political objectives in the international arena, and that they should be supportive of a better functioning global food system.

Expressed another way, the objectives of a harmonious triad of farm, food and trade policies may be said to be maximizing the contribution of the agricultural and food system to national economic growth; avoiding involuntary malnutrition; providing adequate supplies of wholesome food to domestic consumers at reasonable and stable prices; assuring fair and stable returns to farmers and those employed in ancillary industries for their labor, management and capital investments; maximizing agriculture's contribution to the balance of payments over time; promoting expanding and mutually advantageous trading relations with other countries; alleviating hunger among the world's poor; strengthening world economic systems; and promoting harmonious relations between states.

Clearly, an overall policy with such a wide reach is composed of many program areas. And, for the sake of ease of discussion, components that are primarily external in their orientation may be handled separately from domestic measures. But, this expository device does violence to the central reality that global food policies, foreign agricultural trade policies and domestic farm and food policies are inextricably interwoven. This final chapter examines briefly the foreign and domestic agendas in food and agricultural matters facing policy makers in the United States and Canada, paying particular attention to similarities and differences in the circumstances and objectives of each country.

THE EXTERNAL AGENDA

Earlier chapters have examined in some detail the relations between North America, the developing world, the other advanced societies, and the socialist countries. In terms of the well-being of the world's people, the most pressing issue is to remove the scourge of hunger from world society. The fact that in the eighth decade of the twentieth century so many millions continue "to fear for their next day's bread" is evidence enough that the world food system is not functioning well, and it is an indictment of policy makers everywhere. Nonetheless, within the limits of what affluent countries can do to alleviate hunger and malnutrition in other lands, the United States and Canada are not performing badly. Their food production capacity is fully engaged; their agricultural development assistance programs are well directed; they have made important improvements in their food aid programs; and they are committed to improving trade opportunities for the LDCs so as to enhance their ability to buy food on world markets. And, to its credit, the United States has taken an important initiative in making a concrete proposal for the creation of a system of world food grain reserves. Canada has not yet signaled its support for the U.S.'s specific reserves proposal, though it too is committed to the task of enhancing global food security. If there is one element that has been missing in North America's part in the attack on world food inadequacy, it is to be found in the disinclination of U.S. authorities to provide U.S. grain growers with adequate protection against the development of short-run surpluses, thereby introducing a risk that U.S. food production might be interrupted. To the extent that "food security begins with farmer security," and having regard for the importance of the United States in world grain production and trade, many view this as a weakness in the world food system of some importance.

Some changes in the world economic order that will favor the LDCs unquestionably are needed, and, in the process, some old economic orthodoxies will have to be reworked. The direction of change seems bound to involve some substitution of planning and political guidance for market forces in shaping future economic relations between the poor countries and the rich. However, in attempting to give momentum to "the first real dialogue of mankind" and effect to the aspirations they harbor, the LDCs seem to have over-reached themselves in the field of international commodity policy. For reasons discussed in Chapter 4, the UNCTAD's integrated program for commodities is repugnant in many of its details, leans too far toward the politicization of the terms of trade, and threatens to diminish both the performance and the prospects of North American agriculture by drawing the industry into an excessively regulatory regime. Again, it has been left to the United States to contest the less plausible of the LDCs' proposals and to advance constructive alternatives.

One of the more important items of unfinished business of the postwar era is to draw agricultural exchanges between the developed countries into a more open and orderly world trading system. The current GATT negotiations present a further occasion for the United States and Canada to press their often repeated but still valid case for a moderation of the protectionist agricultural policies of Europe and Japan. And, the case will have to be pressed with resolution because the importers' preference for autarkic food policies has been strengthened powerfully by recent scarcities and high prices, by questions about the supply price of expanded North American exports and by uncertainties about their unimpeded access to North American supplies in periods of food scarcity. The prospects for significant trade liberalization do not appear promising, and consideration needs to be given to the consequences of the MTNs yielding but meager fruits for North American agriculture.

The dominant relationship between Canada and the United States in facing Europe and Japan on agricultural matters is the identity of their interest in securing improved access to those markets. However, there appears to be an important difference between them on the matter of an intergovernmental commodity agreement for wheat or for all the major grains. U.S. authorities seem determined to avoid an international arrangement for grains that might provide for a price range that was too narrow and inflexible in relation to developments in world grain supply and demand conditions, and a floor price that was so high as to require international market sharing, a sacrifice of the U.S.'s comparative advantage in periods of adequate supply, and a return to costly national budgetary expenditures on production control, stock holding and surplus disposal. By contrast, Canadian authorities appear to view international floor prices for grains, market sharing and global supply management as an extension into the international arena of measures that are of growing prominence in the conduct of national grains policy. That is to say, from a Canadian viewpoint, stabilization of the international wheat market would provide more stable external conditions to complement the recently introduced policy for stabilizing the Prairie grain sector and thereby the economies of the Prairie provinces and of Canadian agriculture as a whole. The difference between the two countries on this crucial matter undoubtedly lies, in part, in the relatively greater dependence of Canada's agricultural, regional and national economies on the production and export of a single commodity, wheat. U.S. agriculture has far greater production diversity and flexibility than the Canadian industry. And, depression in the wheat growing areas of the United States would not strain national unity—the constant preoccupation of Ottawa. However, one suspects that the different approaches rest in part also on the contrasting moods that have recently prevailed in the two countries with respect to agriculture—the one venerating the potency of market forces, the other preoccupied with risk aversion.

Both Canada and the United States derive important economic benefits from expanding their agricultural exchanges with the USSR and other socialist countries. Both face problems arising from the variability of the USSR's grain import requirements. Each country has mechanisms for coping with this feature of East-West agricultural trade—Canada its Wheat Board, the United States a bilateral agreement which permits a measure of political control of its volume. Neither arrangement adds much to stability in the remainder of the world market for grains, and the assured access that the USSR now has to U.S. grain supplies may have destabilizing effects on trade in wheat between Canada and the USSR. Further, the probable nonparticipation of the USSR in an international system of grain reserves would tend to preclude the creation of the one measure that held promise of enhancing world food security and world price stability. The destabilizing influence of the USSR on world food markets is a global food systems problem that remains to be solved. Workable solutions to this problem are not, alas, immediately apparent.

THE DOMESTIC AGENDA

The support of farmers' incomes is not, in present circumstances, an adequate basis in itself for guiding national policy making for the agriculture and food system. Agricultural adjustment, improved overall market conditions and the dwindling political influence of farm people have so changed the circumstances of commercial farmers that their continuous subsidization is neither necessary nor politically feasible. But, the more important change that has occurred is that the policy agenda for food and agriculture has broadened to embrace a wider range of objectives. In particular, the wider concerns of policy makers are to devise food policies that are supportive of the goals of stabilizing the general economy, maintaining balance in external payments and promoting harmonious foreign economic and political relations.

The adequacy and stability of farmers' prices and incomes have not ceased to be important or legitimate objectives for national food policies. It is simply that price stability has replaced income inadequacy as the primary concern of programs for farmers. Further, the influence of unstable food prices on economic stability and on the distribution of incomes in the general economy has joined farm price and income instability as objectives of national policy. Similarly, foreign agricultural trade policy no longer is concerned solely with improving market opportunities for farmers but also with serving macroeconomic goals such as balance in external payments, and with enhancing national influence on the development of international economic relations and systems as they pertain to food and agriculture.

This broadening of the policy agenda for food and agriculture has not been well understood by farmers. They have resented the fact that the influence of farmers' organizations and departments of agriculture on food policy formation has been diluted by the "intrusion" of consumers and a wider range of government departments. They have felt that their pivotal position in the system has not been acknowledged nor their needs adequately recognized. And, there is no doubt that the immediate economic interests of some groups within the farm sector have been affected adversely by particular food policy decisions.

This is inevitable, for a national food policy must be multi-dimensional in its objectives, constituencies and instruments, and its operation necessarily involves trade-offs between conflicting interests given that the prevailing circumstance with which it must deal is one of instability in national and world food markets.

This last point is crucial. National and world food markets have been highly volatile since 1972, and this instability seems likely to prevail for some time to come. Instability in farm product (and factor) markets gives rise to most of the problems with which contemporary and future national food policies must deal. Instabilities in output and demand impede efficient resource use in agriculture and hamper its growth; effect large and arbitrary income transfers within agriculture and between farmers and consumers; feed into food prices and contribute to the "inflationary ratchet"; cause variations in foreign exchange earnings in the short run; depress foreign commercial demand in the long run; complicate the tasks of building a more open and orderly world trading system for agricultural products; and hinder attempts to alleviate hunger around the world.

In these circumstances, food policy must be capable of dealing flexibly with shortage and surplus situations arising at unpredictable intervals. In the short term, food policy must be capable of balancing the ofttimes conflicting interests of producers of crops and of livestock products, of farmers and consumers and of domestic and foreign buyers. A further important trade-off that has to be made is between short-run domestic price stability on the one hand and the desire to increase foreign exchange earnings from food exports on the other. Implicit in all the above is the conflict between the wish to minimize the role of government in and the size of public expenditures on agriculture and the need for governments to modify the influence of market forces.

The debate in the United States and Canada about food policy does not center on the instruments of policy which are required. Indeed, a complete armory of public programs already exists in each country which could be used to meet every conceivable domestic and international objective for food and agriculture. Thus, the issue is not that of filling important policy gaps by enacting new programs. Rather, the controversy focuses on the balance that should be struck between the interests of different groups and on the use that should be made of particular policy instruments. Essentially, this boils down to differences about the extent to which governments should intervene to provide farmers with some insurance against the risks of short-run weaknesses in markets and to provide consumers with protection from high prices in periods of shortages and the appropriate instruments to be used in each case. The matters on which differences have been drawn most sharply are price supports, grain reserves and export controls.

Minimum Price Supports

The arguments in favor of providing farmers with protection against price collapse hinge on issues of efficiency and equity. It is argued that farmers will be unwilling and unable to make the investments in cost-reducing technologies and in the expansion of production unless they are assured of floor prices that give them some protection against severe losses in low price years. Further, equity considerations dictate that because food prices can be held down only by increased agricultural production, society as a whole should share with farmers the risks of expanding supplies and, more especially, the downside risk of poor prices in years of high output and/or curtailed export demand.

U.S. proponents of this view have argued that current target prices for grains, soybeans and cotton are so far below present costs of production that they give farmers no practical protection against financial distress in years of low market prices and have urged that they be raised. Equally, they would like to see producers of livestock products assured of less volatility in feed grain costs. To the end of its term in January 1977, the Nixon-Ford Administration resisted such proposals, arguing that there was a danger of stop-loss price floors for crops becoming inflexible cost of production-based price supports, resulting in the need for high government expenditures on income transfers, production control, stock acquisition, export subsidies, and similar actions. It was argued further that livestock industries could accommodate to instabilities in markets and costs without public assistance. Major farm organizations supported that stance, as did individual producers with strong financial structures.

In Canada, the federal government has moved decisively to share the risks of price collapse in an era of potentially volatile markets, rapid cost changes, higher farm indebtedness, and rigid cost structures. As mentioned in Chapter 2, this has been done for Prairie grains by the government contributing on a two-for-one basis to the Prairie grains receipts stabilization plan, and also by guaranteeing higher minimum support prices for grains than the U.S. target prices. In addition, higher minimum prices for corn and soybeans are provided to Canadian farmers under the revised Agricultural Stabilization Act than are available to U.S. producers. And, minimum prices (with cost escalators) are available to livestock producers under the latter Act and/or through the control over prices exercised by national and provincial producers' marketing boards. Animal agriculture in the United States has no comparable support.

How durable the current U.S. policy will prove to be under the Carter Administration remains to be seen. To date, the political process has led to a rejection of minimum price supports that have practical relevance to the risks faced by grain and soybean producers, and only modest protection has been provided for the livestock sector, albeit that milk, meats and eggs account for 45 percent of gross U.S. farm cash receipts. Should world grain markets weaken, it is conceivable that current U.S. policies would quickly change.

The concern in Canada is not over whether government is doing too little for farmers in the area of risk sharing but whether it is doing too much. Fears have been expressed that the momentum of federal and provincial stabilization schemes and the operations of marketing boards are such that permanent farm income assurance has replaced short-run price stabilization as the objective of policy. A related complaint is that too little regard is being paid to the interests of consumers.

Reserves

The case for an explicit policy to hold national grain reserves as a necessary component of food policy in an era of volatile grain prices has several strands. It is argued that the availability of reserves in years of low output, large exports and high prices would help stabilize food prices and thereby the general price level; avoid large feed grain price increases and violent adjustments in the livestock sector; minimize income redistributions within agriculture, within national society and internationally that are generally regressive and serve no useful economic purpose; help assure the availability of food aid supplies; obviate the need for export controls; and lend more credibility than is possible with a "bare-shelves" policy to attempts by the United States and Canada to persuade foreign countries that they safely can maintain their dependence on North America for part of their food requirements. U.S. analysts and advocates who find these considerations weighty generally have favored the establishment of a formal national buffer stock to be used to contain grain prices within a desired range, with the stock acquisition price being linked with the minimum market price guarantees accorded farmers. Endless effort has been devoted to exploring institutional aspects of such an arrangement and in analyzing stock acquisition and operating costs, the optimum size of stocks, the decision rules governing stock procurement and release, and similar matters. Much of the analytical work has been inconclusive because, with numerous potential target variables to be stabilized, there can be no single optimum stock policy. In addition, the determination and evaluation of distributional aspects of benefits and costs have proved elusive. And, researchers have had no information on such key variables as the response of output and demand to a stabilized environment, the effect of the existence of public stocks on the inventory policies of private firms, and the relative weights attached by various groups to the stability and levels of prices, costs, incomes, and expenditures.

Opponents of public stocking policies have been quick to point out these and other analytical deficiencies and to add that few analyses demonstrate a favorable ratio between the measurable costs and benefits of a stocking operation. Other weighty arguments have been mobilized. It has been pointed out that reducing the amplitude of fluctuations of grain prices would have a barely discernible impact on the consumer price index; that livestock industries were subject to violent cyclical instabilities even during the long period when large public grain stocks existed and feed grain prices were low and stable;

and that importing countries showed no disposition to embrace freer trade policies for grains when North America's grain stocks were huge.

To date, the U.S. government, and U.S. grain growers generally, have opposed the creation of national grain reserves for additional reasons. The Nixon-Ford Administration wished to avoid the financial and other consequences of "getting the government back into the grain business," and it feared that importing countries would not pursue the U.S. proposal for creating a world food security reserve (or generally accept more responsibility for stock holding) if the major exporters took unilateral action to rebuild their grain stockpiles. Naturally enough, grain growers have not desired a mechanism that would lower price peaks. Some have also claimed that the long-run equilibrium price would be lowered by the stocks "overhanging" the market, though this is not a necessary result, for it derives from a fallacious correlation between large grain stocks and low grain prices in earlier years; neglects the fact that public stocks could be more firmly held than private stocks; and gives little weight to the consideration that decision rules on release prices could readily be devised that would obviate this effect.

In contrast to the United States, Canada is in the odd position of having an explicit policy on holding national grain reserves, the details of which are not in the public domain. That is to say, the Canadian Wheat Board, which is a public agency, is known to manage its inventories of grain in a systematic manner, but it does not make public either its stock objectives or its decision rules.

If the U.S. proposal for the establishment of a multilateral system of grain reserves—which in principle is supported by Canada—is accepted, U.S. and Canadian authorities will perforce have to acquire publicly financed grain reserves. If the proposal does not command international support, a decision will have to be made on the creation of national reserve stocks. One might anticipate that two considerations will command particular attention. First, long-run export demand for North American grains likely will be greater if public stocks are held than if they are not. The possession of reserves will not guarantee that importing countries will sustain their dependence on imported supplies, but their nonexistence will surely strengthen the impulse in these countries to increase their self-sufficiency and their efforts to diversify supply sources. Second, reserves may be preferable to their most probable alternative, the limitation of exports.

Export Controls

Both the United States and Canada have been forced to resort to the limitation of grain export shipments at various times since 1973. In Canada, this is effected readily and unobtrusively by the Canadian Wheat Board's having a monopoly in the marketing of wheat which permits it to allocate supplies between domestic uses and foreign sales.

In the United States, regulatory intervention has been effected by executive action and under the terms of the U.S.-USSR grains agreement.

There is a feeling in both countries that this is a policy instrument that should go unused except in the most exceptional circumstances, as when poor crops and/or heavy export demand threaten to raise domestic food and feed prices to politically intolerable levels, when sporadic buying threatens the availability of supplies to regular customers, or if it should be necessary to deny benefits to nonparticipants in a multilateral reserves scheme.

However, the controversy is not whether they should be used sparingly and responsibly (and in strict accordance with any new code governing access to supplies that may be written into the GATT), but whether they should be used at all. Grain growers feel strongly that they should not. And, many others consider that it would be a better example for the world trading system, inspire more confidence in the United States and Canada as regular suppliers and avoid difficult choices as to which buyers to supply and which to deny, if export restrictions were rendered unnecessary by the creation of adequate national reserves. It scarcely seems possible that national economic stability can be achieved with neither reserves nor export controls.

CONCLUSIONS

One is impressed by the similarities in the problems faced by the United States and Canada in developing agricultural and food policies appropriate to their position in an interdependent world food economy characterized by instability and uncertainty. Both countries face the task of evolving policies that will maximize the contribution of their agricultural sectors to national economic growth, to internal price stability and to external payments balance, and that will permit them to strengthen their influence on world food systems, international trading arrangements and international economic relations generally.

In the international arena, there is a broad communality of purpose, with the only major difference being in the greater importance that Canada attaches to an international stabilization arrangement for grains to complement its domestic stabilization programs for this vital sector. Neither country has moved explicitly to prepare the noncompetitive sectors of its agriculture to withstand the intensified competition in its domestic markets that will result from trade liberalization. One notes, too, that most of the initiatives on or reaction to proposals for reshaping international trading arrangements for agricultural products and strengthening the world food systems have come from the United States rather than from Canada.

At home, both countries are moving perceptibly from a narrow preoccupation with domestic farm policy toward an integrated approach to decision making with respect to global food systems problems, foreign agricultural trade policies, national food policies, and domestic farm programs. Each country is seeking to evolve a set of mutually supportive policies in each of these areas. Both are experiencing difficulties in this task, in part because the existing structure of government is not well adapted to decision making on the multifaceted and interrelated international and national aspects of a comprehensive food policy, but mainly because there are difficult choices to be made between more numerous competing interest groups and objectives.

It is apparent that in adapting domestic farm programs to a situation of greater uncertainty and instability in world and national food markets, in making farm policy a component of food policy and in attempting to make food policy better serve national macroeconomic and foreign economic policy, each country has chosen different nodal points for emphasis, opted for differing degrees of government involvement and used different policy instruments. It is here that the most worrisome aspect of the U.S.-Canadian relationship on agricultural matters is to be found.

In the United States, the ascendant assumption is that market forces should play a greater role in the domestic and international dimensions of agricultural markets. But, the thrust of such interventions as remain recently has been to favor the interests of consumers over those of farmers. In Canada, by contrast, the direction of change has been to extend government involvement in the agricultural and food system, and these interventions have been heavily oriented to farmer interests, particularly to strengthening systems for providing farmers with minimum returns. The commodity coverage of minimum price support arrangements is wider in Canada than in the United States, the levels of support accorded common commodities is higher, and Canada uses different instruments of price support from those employed in the United States.

These differences have already led to important bilateral conflicts in cross-border trade. To date, these disputes have been confined to trade in livestock products. The potential for conflicts in this area could be extended in the future if Canada were to adopt a two-price system for feed grains in order to give further assistance to its livestock industries. Such a step has been mooted. In addition, the scope for disputes would widen if Canada were to extend its commitment to stabilizing farmers' returns to a wider range of products, move from the objective of stabilization to that of support of farmers' incomes, and provide such assurances by the regulation of marketed supplies rather than by direct payments.

The danger of bilateral trade disputes is inherent in the present instability within world agriculture and in the different policy priorities of the two countries. It is difficult to see how they can be avoided, though they might be mitigated by more explicit consideration of the effects of national policies on the partner and earlier consultations on policy changes that are under consideration. In addition, a reduction in the rate of inflation would ease the pressure on farmers' costs and thereby their need for programs of income protection. Furthermore, greater stability in world agricultural markets would extend to the continental market and narrow the scope for the adoption of conflicting policies on the two sides of the common border.

MEMBERS OF THE CANADIAN-AMERICAN COMMITTEE

Co-Chairmen

ROBERT M. MacINTOSH
Executive Vice-President, The Bank of Nova Scotia, Toronto, Ontario

RICHARD J. SCHMEELK
Partner, Salomon Brothers, New York, New York

Members

I.W. ABEL
President, United Steelworkers of America, AFL-CIO, Pittsburgh, Pennsylvania

JOHN N. ABELL
Vice President and Director, Wood Gundy Limited, Toronto, Ontario

R.L. ADAMS
Executive Vice President, Continental Oil Company, Stamford, Connecticut

J.D. ALLAN
President, The Steel Company of Canada, Limited, Toronto, Ontario

J.A. ARMSTRONG
President and Chief Executive Officer, Imperial Oil Limited, Toronto, Ontario

IAN A. BARCLAY
Chairman, British Columbia Forest Products Limited, Vancouver, British Columbia

MICHEL BELANGER
President and Chief Executive Officer, Provincial Bank of Canada, Montreal, Quebec

ROY F. BENNETT
President and Chief Executive Officer, Ford Motor Company of Canada, Limited, Oakville, Ontario

ROD J. BILODEAU
Chairman of the Board, Honeywell Limited, Scarborough, Ontario

ROBERT BLAIR
President and Chief Executive Officer, Alberta Gas Trunk Line Company Limited, Calgary, Alberta

J.E. BRENT
Director, IBM Canada Ltd., Toronto, Ontario

PHILIP BRIGGS
Senior Vice President, Metropolitan Life Insurance Company, New York, New York

ARDEN BURBIDGE
Park River, North Dakota

NICHOLAS J. CAMPBELL, JR.
New York, New York

SHIRLEY CARR
Executive Vice-President, Canadian Labour Congress, Ottawa, Ontario

W.R. CLERIHUE
Executive Vice-President, Staff and Administration, Celanese Corporation, New York, New York

HON. JOHN V. CLYNE
MacMillan Bloedel Limited, Vancouver, British Columbia

STANTON R. COOK
President, Tribune Company, Chicago, Illinois

THOMAS E. COVEL
Marion, Massachusetts

GEORGE B. CURRIE
Vancouver, British Columbia

J.S. DEWAR
President, Union Carbide Canada Limited, Toronto, Ontario

JOHN H. DICKEY
President, Nova Scotia Pulp Limited, Halifax, Nova Scotia

JOHN S. DICKEY
President Emeritus and Bicentennial Professor of Public Affairs, Dartmouth College, Hanover, New Hampshire

THOMAS W. diZEREGA
Vice President, Northwest Pipeline Corporation, Salt Lake City, Utah

WILLIAM DODGE
Ottawa, Ontario

WILLIAM EBERLE
President and Chief Executive Officer, Motor Vehicle Manufacturers Association of the United States, Washington, D.C.

STEPHEN C. EYRE
Comptroller, Citibank, N.A., New York, New York

A.J. FISHER
President, Fiberglas Canada Limited, Toronto, Ontario

CHARLES F. FOGARTY
Chairman and Chief Executive Officer, Texasgulf Inc., New York, New York

ROBERT M. FOWLER
President, C.D. Howe Research Institute, Montreal, Quebec

JOHN F. GALLAGHER
Vice President International Operations, Sears, Roebuck and Company, Chicago, Illinois

90

W.D.H. GARDINER
Deputy Chairman & Executive Vice President, The Royal Bank of Canada, Toronto, Ontario

CARL J. GILBERT
Dover, Massachusetts

DONALD R. GRANGAARD
President, First Bank System, Inc., Minneapolis, Minnesota

PAT GREATHOUSE
Vice President, International Union, UAW, Detroit, Michigan

A.D. HAMILTON
President and Chief Executive Officer, Domtar Limited, Montreal, Quebec

JOHN A. HANNAH
Executive Director, World Food Council, New York, New York

ROBERT H. HANSEN
Senior Vice President—International, Avon Products, Inc., New York, New York

G.L. HARROLD
President, Alberta Wheat Pool, Calgary, Alberta

F. PEAVEY HEFFELFINGER
Director Emeritus, Peavey Company, Minneapolis, Minnesota

R.A. IRWIN
Chairman, Consolidated-Bathurst Limited, Montreal, Quebec

ROBERT H. JONES
President, The Investors Group, Winnipeg, Manitoba

EDGAR F. KAISER, JR.
President and Chief Executive Officer, Kaiser Resources Ltd., Vancouver, British Columbia

JOSEPH D. KEENAN
Washington, D.C.

DONALD P. KELLY
President and Chief Operating Officer, Esmark, Inc., Chicago, Illinois

DAVID KIRK
Executive Secretary, The Canadian Federation of Agriculture, Ottawa, Ontario

LANE KIRKLAND
Secretary-Treasurer, AFL-CIO, Washington, D.C.

WILLIAM J. KUHFUSS
Mackinaw, Illinois

J.L. KUHN
President and General Manager, 3M Canada Limited, London, Ontario

HERBERT H. LANK
Director, Du Pont of Canada Limited, Montreal, Quebec

PAUL LEMAN
President, Alcan Aluminum Limited, Montreal, Quebec

EDMOND A. LEMIEUX
General Manager, Finance, Hydro-Quebec, Montreal, Quebec

RICHARD A. LENON
President, International Minerals & Chemical Corporation, Libertyville, Illinois

FRANKLIN A. LINDSAY
Chairman, Itek Corporation, Lexington, Massachusetts

*L.K. LODGE
President and Chairman, IBM Canada Ltd., Don Mills, Ontario

L.H. LORRAIN
President, Canadian Paperworkers Union, Montreal, Quebec

M.W. MACKENZIE
Vice Chairman, Canron Limited, Montreal, Quebec

WILLIAM MAHONEY
National Director, United Steelworkers of America, AFL-CIO-CLC, Toronto, Ontario

JULIEN MAJOR
Executive Vice-President, Canadian Labour Congress, Ottawa, Ontario

PAUL M. MARSHALL
Calgary, Alberta

FRANCIS L. MASON
Senior Vice President, The Chase Manhattan Bank, New York, New York

DENNIS McDERMOTT
UAW International Vice President and Director for Canada, International Union, United Automobile, Aerospace and Agricultural Implement Workers of America, Willowdale, Ontario

WILLIAM J. McDONOUGH
Executive Vice President, The First National Bank of Chicago, Chicago, Illinois

WILLIAM C.Y. McGREGOR
International Vice President, Brotherhood of Railway, Airline and Steamship Clerks, Montreal, Quebec

H. WALLACE MERRYMAN
Chairman and Chief Executive Officer, Avco Financial Services, Inc., Newport Beach, California

JOHN MILLER
President, National Planning Association, Washington, D.C.

COLMAN M. MOCKLER, JR.
Chairman and President, The Gillette Company, Boston, Massachusetts

DONALD R. MONTGOMERY
Secretary-Treasurer, Canadian Labour Congress, Ottawa, Ontario

JOSEPH MORRIS
President, Canadian Labour Congress, Ottawa, Ontario

---

*Became a member of the Committee after the Statement was circulated for signature.

RICHARD W. MUZZY
Group Vice President-International, Owens-Corning Fiberglas Corporation, Toledo, Ohio

THEODORE NELSON
Executive Vice President, Mobil Oil Corporation, New York, New York

THOMAS S. NICHOLS, SR.
Director, Olin Corporation, New York, New York

CARL E. NICKELS, JR.
Senior Vice President, Finance and Law, The Hanna Mining Company, Cleveland, Ohio

JOSEPH E. NOLAN
Tacoma, Washington

HON. VICTOR deB. OLAND
Halifax, Nova Scotia

CHARLES PERRAULT
President, Perconsult Ltd., Montreal, Quebec

RICHARD H. PETERSON
Chairman of the Board, Pacific Gas and Electric Company, San Francisco, California

R.J. RICHARDSON
Vice President and Treasurer, E.I. du Pont de Nemours & Co., Inc., Wilmington, Delaware

BEN L. ROUSE
Vice President-Business Machines Group, Burroughs Corporation, Detroit, Michigan

THOMAS W. RUSSELL, JR.
New York, New York

A.E. SAFARIAN
Dean, School of Graduate Studies, University of Toronto, Toronto, Ontario

W.B. SAUNDERS
Group Vice President, Cargill, Incorporated, Minneapolis, Minnesota

ARTHUR R. SEDER, JR.
Chairman and President, American Natural Resources Company, Detroit, Michigan

A.R. SLOAN
President and General Manager, Continental Can International, Stamford, Connecticut

R.W. SPARKS
President and Chief Executive Officer, Texaco Canada Limited, Don Mills, Ontario

EDSON W. SPENCER
President and Chief Executive Officer, Honeywell Inc., Minneapolis, Minnesota

W.A. STRAUSS
Chairman and President, Northern Natural Gas Company, Omaha, Nebraska

ROBERT D. STUART, JR.
Chairman, The Quaker Oats Company, Chicago, Illinois

A. McC. SUTHERLAND
Senior Vice President, INCO Limited, Toronto, Ontario

DWIGHT D. TAYLOR
Senior Vice President, Crown Zellerbach Corporation, San Francisco, California

ROBERT B. TAYLOR
Chairman, Ontario Hydro, Toronto, Ontario

WILLIAM I.M. TURNER, JR.
President and Chief Executive Officer, Consolidated-Bathurst Limited, Montreal, Quebec

W.O. TWAITS
Toronto, Ontario

MELVIN J. WERNER
Vice President, Farmers Union Grain Terminal Association, Saint Paul, Minnesota

JOHN R. WHITE
New York, New York

HENRY S. WINGATE
Formerly Chairman and Chief Officer, INCO Limited, New York, New York

WILLIAM W. WINPISINGER
General Vice President, International Association of Machinists and Aerospace Workers, Washington, D.C.

THOMAS WINSHIP
Editor, *The Boston Globe,* Boston, Massachusetts

FRANCIS G. WINSPEAR
Edmonton, Alberta

D. MICHAEL WINTON
Chairman, Pas Lumber Company Limited, Minneapolis, Minnesota

GEORGE W. WOODS
President, TransCanada Pipelines, Toronto, Ontario

WILLIAM S. WOODSIDE
President, American Can Company, Greenwich, Connecticut

*DON WOODWARD
International Trade Affairs Representative, National Association of Wheat Growers, Pendleton, Oregon

ADAM H. ZIMMERMAN
Executive Vice President, Noranda Mines Limited, Toronto, Ontario

Honorary Members

HON. N.A.M. MacKENZIE
Vancouver, British Columbia

HAROLD SWEATT
Honorary Chairman of the Board, Honeywell Inc., Minneapolis, Minnesota

DAVID J. WINTON
Minneapolis, Minnesota

---

*Became a member of the Committee after the Statement was circulated for signature.

SELECTED PUBLICATIONS OF THE CANADIAN-AMERICAN COMMITTEE*

Commercial Relations

CAC-40 *Industrial Incentive Policies and Programs in the Canadian-American Context,* by John Volpe. 1976 ($2.50)

CAC-38 *A Balance of Payments Handbook,* by Caroline Pestieau. 1974 ($2.00)

CAC-32 *Toward a More Realistic Appraisal of the Automotive Agreement,* a Statement by the Committee. 1970 ($1.00)

CAC-31 *The Canada-U.S. Automotive Agreement: An Evaluation,* by Carl E. Beigie. 1970 ($3.00)

CAC-25 *A New Trade Strategy for Canada and the United States,* a Statement by the Committee. 1966 ($1.00)

Energy and Other Resources

CAC-41 *Coal and Canada-U.S. Energy Relations,* by Richard L. Gordon. 1976 ($3.00)

CAC-39 *Keeping Options Open in Canada-U.S. Oil and Natural Gas Trade,* a Statement by the Committee. 1975 ($1.00)

CAC-37 *Canada, the United States, and the Third Law of the Sea Conference,* by R. M. Logan. 1974 ($3.00)

CAC-36 *Energy from the Arctic: Facts and Issues,* by Judith Maxwell. 1973 ($4.00)

Investment

CAC-33 *Canada's Experience with Fixed and Flexible Exchange Rates in a North American Capital Market,* by Robert M. Dunn, Jr. 1971 ($2.00)

CAC-29 *The Performance of Foreign-Owned Firms in Canada,* by A. E. Safarian. 1969 ($2.00)

CAC-24 *Capital Flows Between Canada and the United States,* by Irving Brecher. 1965 ($2.00)

Other

CAC-43 *Agriculture in an Interdependent World: U.S. and Canadian Perspectives,* by T. K. Warley. 1977 ($4.00)

CAC-42 *A Time of Difficult Transitions: Canada-U.S. Relations in 1976,* a Staff Report. 1976 ($2.00)

CAC-35 *The New Environment for Canadian-American Relations,* a Statement by the Committee. 1972 ($1.50)

CAC-30 *North American Agriculture in a New World,* by J. Price Gittinger. 1970 ($2.00)

---

*These and other Committee publications may be ordered from the Committee's offices at 2064 Sun Life Building, Montreal, Quebec H3B 2X7, and at 1606 New Hampshire Avenue, N.W., Washington, D.C. 20009. Quantity discounts are given. A descriptive flyer of these publications is also available.